Michael Wilson
PO BOX 4424
CLARKSVILLE, TN 37044

wilsongis@gmail.com

Learning ArcGIS 10.2 Basics

By Michael Wilson

Contents

Foreword

This Book...

The purpose of this book is to introduce the reader to Geographic Information Systems (GIS) and ArcGIS Desktop 10.1 or 10.2 by ESRI.

Over the years, I have worked with various GIS lessons, tutorials and how-to books, websites, and other resources. This book is a compilation of all the various "best practices" that I have learned. My goal is to help those new to GIS and ArcGIS to develop a solid foundation upon which to build.

Required Software

The lessons in this book require access to ArcGIS 10.1 or 10.2.

Real World Application (RWA)

While writing and assembling this workbook, I have tried to keep in mind that "book learning" is often different than how something gets done in the "real world". In each lesson of the following sections, I have included information on "Real World Application" or "RWA". The RWA information is meant to tie the various exercises to how the skill would typically be used in a commercial, business, or academic GIS production environment. My intent is to bridge the gap between book learning and practical application.

Special Thanks

I would like to take a moment to thank the various friends that have assisted by proofreading my work. I would especially like to thank Andrew Shepard-Smith who went above and beyond with his great edits and suggestions.

Lastly, I would like to thank my family for their love and support. I owe a special thanks to my wife, Tara, whom is my source of inspiration. I love you.

The Data

The Data for these lessons can be downloaded from http://www.arcbasics.com/data

All Data - http://www.arcbasics.com/data/ArcBasics.zip

Imagery - http://www.arcbasics.com/data/Images.zip

Introduction - http://www.arcbasics.com/data/Intro.zip

The GIS Workspace - http://www.arcbasics.com/data/Workspace.zip

Displaying Spatial Data - http://www.arcbasics.com/data/DSD.zip

Selecting Features - http://www.arcbasics.com/data/Select.zip

Symbolizing Data - http://www.arcbasics.com/data/Symbols.zip

Making A Layout - http://www.arcbasics.com/data/Layout.zip

Mid-Term Exercise 1 - http://www.arcbasics.com/data/Mid-Term1.zip

Using Attribute Data - http://www.arcbasics.com/data/UAD.zip

Creating and Editing Data - http://www.arcbasics.com/data/CED.zip

Coordinate system and Projection - http://www.arcbasics.com/data/Coord.zip

Geoprocessing - http://www.arcbasics.com/data/GeoP.zip

Final Exercise - http://www.arcbasics.com/data/Final.zip

Setting up the Data

1. Create a folder on your hard drive called "ArcBasics". Be sure to note the location of the folder.
2. Download and extract the data files into the "ArcBasics" folder.
3. The data is now ready to be utilized.

Chapter 1 - Introduction

Introduction

As you build any structure, it is important to begin with a solid foundation. In this section, you will learn introductory concepts, principles, and terminology related to GIS and ESRI's ArcGIS software suite.

<u>What is GIS?</u>

A Geographic Information System (GIS) integrates hardware, software, and data for capturing, managing, analyzing, and displaying all forms of geographically referenced information.

In simpler terms:

"GIS is not about making maps, per se. – It is about analyzing often large sets of data to generate information – hypotheses, conclusions, insights, new hunches – about widely varied socio-economic phenomena... It is about telling a story..."

<u>J. T. Johson – GIS as a unifying methodology in Journalism</u> {J. T. Johnson, 2003}

What is ArcGIS?

ArcGIS

Parcel editing with ArcMap 10.1 on Windows 7

Developer(s)	Esri
Initial release	December 27, 1999; 13 years ago
Stable release	10.1 / June 11, 2012; 10 months ago
Development status	Active
Operating system	**Desktop**: Windows XP SP2 and later, Windows Server 2003 SP2 and later;[1] **Server** (x64 only) additionally supports: RHEL 5 and later, SLES 11 and later;[2] **Mobile**: iOS 3.1.2 and later, Android 2.2 and later, Windows Phone 7 and later,[3] Windows Mobile 6 and later[4]
Available in	English, Arabic, French, German, Italian, Japanese, Portuguese, Russian, Simplified Chinese, Spanish[5]
Type	Geographic information system
License	Proprietary commercial software
Website	http://www.esri.com/software/arcgis/

Figure 1: Wikipedia ArcGIS Summary[1]

From Wikipedia[2]:

ESRI's ArcGIS is a GIS for working with maps and geographic information. It is used for: creating and using maps; compiling geographic data; analyzing mapped information; sharing and discovering geographic information; using maps and geographic information in a range of applications; and managing geographic information in a database.

#

Intro Objectives

Figure 2: ArcMap Launch

This section will help you understand:
- The ArcGIS Desktop Suite
    - License Levels
    - ArcGIS Components
    - ArcGIS extensions
- GIS Data Types
- ESRI Data formats
- Navigating in ArcGIS
    - ArcCatalog
    - ArcMap

    ☐   ArcToolbox
- Getting Help

#

The ArcGIS Desktop Suite

The ArcGIS suite is made up of three different license levels (Figure 3). The levels are:

- ArcGIS for Desktop Basic (formerly known as ArcView)
  - ☐ Map Creation
  - ☐ Interactive Visualization
- ArcGIS for Desktop Standard (formerly known as ArcEditor)
  - ☐ Map Creation
  - ☐ Interactive Visualization
  - ☐ Multiuser Editing
  - ☐ Advanced Data Management
- ArcGIS for Desktop Advanced (formerly known as ArcInfo)
  - ☐ Map Creation and
  - ☐ Interactive Visualization
  - ☐ Multiuser Editing
  - ☐ Advanced Data Management
  - ☐ Advanced Analysis,
  - ☐ High-End Cartography
  - ☐ Extensive Database Management

	Basic (ArcView)	Standard (ArcEditor)	Advanced (ArcInfo)
Map Creation and Interactive Visualization			
Visually model and spatially analyze a process or workflow.	✓	✓	✓
Create interactive maps from file, database, and online sources.	✓	✓	✓
Create street-level maps that incorporate GPS locations.	✓	✓	✓
View CAD data or satellite images.	✓	✓	✓
Generate reports and charts.	✓	✓	✓
Multiuser Editing and Advanced Data Management			
Complete GIS data editing capabilities.		✓	✓
Edit a multiuser enterprise geodatabase.		✓	✓
Use disconnected editing in the field.		✓	✓
Store historical snapshots of your data.		✓	✓
Automate quality control.		✓	✓
Create spatial data from scanned maps.		✓	✓
Use raster-to-vector conversion.		✓	✓
Advanced Analysis, High-End Cartography, and Extensive Database Management			
Advanced GIS data analysis and modeling			✓
Atlaslike, publication-quality maps			✓
Advanced data translation and creation			✓
Advanced feature manipulation and processing			✓
Data conversion for CAD, raster, dBASE, and coverage formats			✓

Figure 3: Overview of License Levels[3]

For more information, see ESRI's Website -
http://www.esri.com/software/arcgis/about/gis-for-me

\#

ArcGIS Components

Regardless if you are working with ArcGIS Basic, Standard or Advanced, the software is made of three distinct components, ArcCatalog, ArcMap, and ArcToolbox.

- ArcCatalog (Figure 4) is an interface that allows users to access, interact, and manage spatial data. This data can be stored in local folders, databases, and in geodatabases that are part of an enterprise database. Similar to Microsoft Explorer, ArcCatalog allows users to perform a variety of file functions such as rename, copy, delete, and preview. ArcCatalog also provides an interface to various datasets metadata. The metadata can be created or read directly from ArcCatalog.

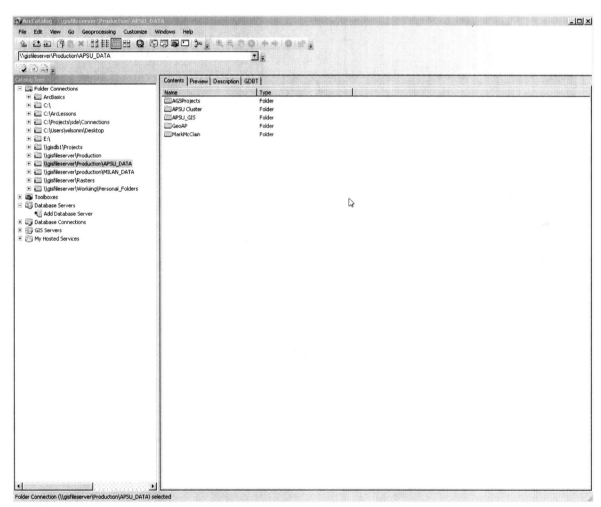

Figure 4: ArcCatalog

- ArcMap (Figure 5) allows users to directly manipulate spatial information. Users have the ability to display, query, analyze, and create high quality printed maps. ArcMap is both a data viewing platform and a powerful tool for editing spatial data.

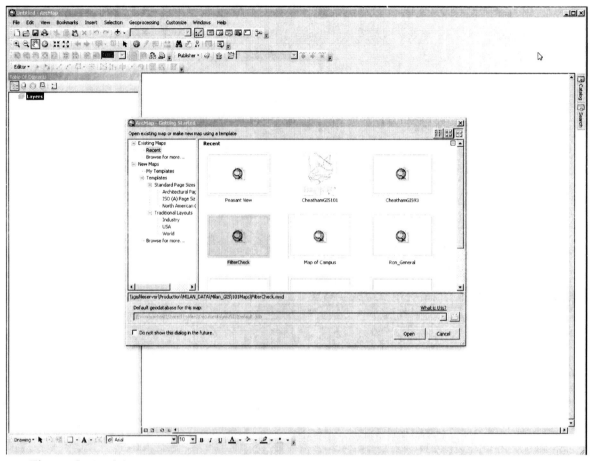

Figure 5: ArcMap

- ArcToolbox (Figure 6) is an interface for accessing and organizing a collection of geoprocessing tools, models and scripts. In ArcGIS versions prior to ArcGIS 10, ArcToolbox contained search, index, and results tabs. These tabs have been replaced by the Search window (Figure 7) and the Results window (Figure 8), available from the Geoprocessing menu.

Figure 6: ArcToolbox

Figure 7: The Search Window

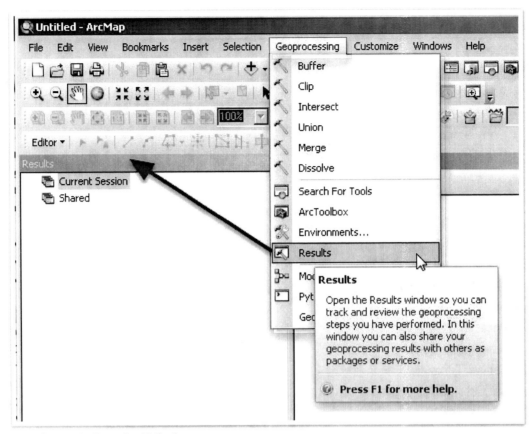

Figure 8: Results Window

#

ArcGIS Extensions

ESRI and a variety of third-party developers offer a wide variety of extensions that greatly increase the capabilities of ArcGIS. The ESRI extensions typically fall into two categories: analysis and productivity (Figure 9). Some examples of analysis extensions include Spatial Analyst, 3D Analyst, and Network Analyst to name a few. Productivity extensions include ArcGIS Publisher, Maplex, Data Reviewer, and others.

Extensions

Add more capabilities to ArcGIS for Desktop with extensions. These specialized tools allow you to perform more sophisticated tasks such as raster geoprocessing and 3D analysis.

Analysis	Key Benefits
ArcGIS 3D Analyst	Analyze your data in a realistic perspective.
ArcGIS Geostatistical Analyst	Use advanced statistical tools to investigate your data.
ArcGIS Network Analyst	Perform sophisticated routing, closest facility, and service area analysis.
ArcGIS Schematics	Represent and understand your networks to shorten decision cycles.
ArcGIS Spatial Analyst	Derive answers from your data using advanced spatial analysis.
ArcGIS Tracking Analyst	Reveal and analyze time-based patterns and trends in your data.
Business Analyst Online Reports Add-In	Directly access demographic reports and data from Business Analyst Online (BAO) for trade areas and sites created in the desktop.

Productivity	Key Benefits
ArcGIS Data Interoperability	Eliminate barriers to data use and distribution.
ArcGIS Data Reviewer	Automate, simplify, and improve data quality control management.
ArcGIS Publisher	Freely share your maps and data with a wide range of users.
ArcGIS Workflow Manager*	Better manage GIS tasks and resources.
ArcScan for ArcGIS**	Increase efficiency and speed up raster-to-vector data conversion time.
Maplex for ArcGIS**	Create maps that communicate more clearly with automatically positioned text and labels.

Solution Based	Key Benefits
ArcGIS Defense Solutions (includes ArcGIS Military Analyst, Grid Manager, and MOLE)	Create workflows, processes, and symbology to support defense and intelligence planning.
ArcGIS for Aviation*	Use the full power of GIS to efficiently manage aeronautical and airports information.
ArcGIS for Maritime*	Gain value from bathymetry and leverage nautical data beyond charting.
Esri Defense Mapping***	Efficiently manage defense specification-compliant products.
Esri Production Mapping*	Standardize and optimize your GIS production.
Esri Roads and Highways*	Easily manage, visualize, and analyze transportation networks.

* Requires ArcGIS for Desktop Standard or Advanced

** Starting with the release of ArcGIS 10.1, this extension is included with ArcGIS for Desktop at no extra cost

*** Requires ArcGIS for Desktop Advanced

Note: Unless noted, extensions can be used with ArcGIS for Desktop Basic, Standard, and Advanced.

Figure 9: Overview of Extensions[4]

For more information see the ESRI Website -

http://www.esri.com/software/arcgis/arcgis-for-desktop/extensions

GIS Data Types
Vector Data

The vector data model is used to represent discrete features that are defined as points, lines, and polygons in a geographic information system (GIS). Vector data represent features as pairs of x,y coordinates.

Points are geographic features represented by a single x,y coordinate pair. Points are are zero-dimensional features that have neither length nor area. Example of point data includes cities, wells, schools, and other points of interest.

Lines are formed by connecting two x,y coordinate pairs. When more than two x,y points are used to create a line, the detail of the line will be greater. Lines have length but do not have area. Examples of line data include streets, trails, and railroads.

Polygons are an area fully encompassed by a series of connected lines. Polygons are used to represent areas on the Earth's surface. A polygon contains one type of data such as a lake, city boundary, or landmass. Polygons have a measurable perimeter and area.

Raster Data

The raster data model represents map features as cells in a grid matrix. The matrix is organized by continuous, evenly spaced rows and columns. Each cell is coded with an attribute that represents a data parameter (such as, the predominant type of sediment, the average bottom depth, or even the sea surface temperature) that appears within the cell.

#

Image Formats

Out of the box, ArcMap supports a variety of image formats. Some of the supported formats include:

- BSQ, BIL, BIP
- ERDAS IMAGINE FILES
- Joint Photographic Experts Group (JPEG)
- Bitmap (BMP)
- Mr. SID (Compressed image format from LizardTech[5])

Image data is composed of rasters. The rasters are a grid of rows and columns. Each cell in the grid is a pixel. Each pixel represents an area on the Earth's surface. This area is referred to as the image resolution. For example, when an image has one-foot resolution, each pixel represents a 1' x 1' area on the ground. Spatial imagery is geo-referenced. Geo-referenced data is information that is high to real world coordinates. In short, a pixel in an image represents and has identical coordinates as a specified location on the Earth's surface.

The geo-referenced information is either stored in the image header or in a separate world file. The world file contains coordinate information that allows the image to be displayed properly. This information is used by ArcMap to translate the image to its real world coordinates. A world file always accompanies its associated image. World files use the same name as the image with a file extension that ends in W. For example, an image named APSU.tif would have a world file called APSU.tfw.

Images with header file extension (hdr) include BSQ, BIL, or BIP file from the USGS. This file contains a set of entries that describe the attributes of the data. A .hdr file contains the following information:

- nrows <number of rows>
- ncols <number of columns>
- nbands <number of bands of data (BIL and BIP formats)>
- ulxmap <pixel 1,1 x coordinate (upper-left pixel)>
- ulymap <pixel 1,1 y coordinate (upper-left pixel)>

#

Coordinate Systems

Spatial data are data that are georeferenced. That is, they are referenced to the surface of the Earth using either a geographic coordinate system or a projected coordinate system. Geographic coordinate systems use both latitude and longitude for coordinates. Even though only two coordinates are required to locate a point on the Earth's surface, latitude/longitude are three-dimensional coordinates because the Earth's surface is three-dimensional.

Projected coordinate systems use a mathematical conversion to transform latitude and longitude coordinates that fall on the Earth's three-dimensional surface to a flat two-dimensional surface. A projected coordinate system is made up of a spheroid, datum, projection, and horizontal units (i.e. map units).

ArcGIS can work with data stored in either geographic or projected coordinates (Figure 10).

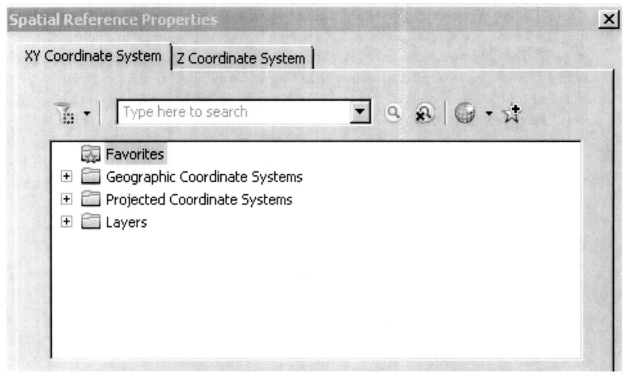

Figure 10: Spatial References Properties Window

For a more detailed explanation of Coordinate Systems, please see the Coordinate System Section.

#

Vector Data File Formats

Shapefile (.shp)

The ESRI Shapefile is a popular geospatial vector data format for GIS data. It has been developed and regulated by ESRI, and has become an industry standard file type for spatial data. A wide variety of GIS software systems can read and write shapefiles including MapInfo, QGIS, GRASS, and uDig.

A shapefile is comprised of at least three core files. The files have the same prefix name and have the following extensions: .shp, .shx, .dbf. There are numerous other optional extensions that may be present. In order to transmit a shapefile to another user, all files with the same prefix name must be sent.

Geodatabases (.mdb and .gdb)

The geodatabase is a collection of spatial datasets and is designed for data storage. It is used by ArcGIS software and managed in either a file folder or a relational database. There are three types of geodatabases: file, personal, and ArcSDE.

Smart Data Compression (SDC)

Smart Data Compression is a compressed GIS dataset format developed by ESRI. It stores all types of feature data and attribute information together as a core data structure.

Layer File (.lyr)

The ESRI layer file does not contain spatial data, but does contain reference to symbology and classification data. A layer file is a reference to a dataset.

ArcInfo Coverage

ESRI ArcInfo Coverage is a georelational data model that stores vector data. The coverage data format is a legacy format.

Raster Data File Formats

ArcGIS can handle many types of raster data. Some of the most popular formats include:

MrSID (.sid)

MrSID (pronounced "mister sid") is a proprietary format of developed by [LizardTech](http://www.lizardtech.com/) to encode georeferenced raster graphics, such as orthophotos. MrSID is an acronym that stands for multi-resolution seamless image database. Orthophotos are photos that have been corrected for changes in elevation on the Earth's surface.

ECW

ECW is a proprietary wavelet compression image format optimized for spatial imagery. It was developed by Earth Resource Mapping (ER Mapper).

JPEG 2000

JPEG 2000 is an image coding system that uses state-of-the-art compression techniques based on wavelet technology. It is a non-proprietary image compression format based on ISO standards, and the standardized filename extension is .jp2.

Based upon information found as:
ESRI - http://www.esri.com
Geospatial Data Formats - http://www.lib.ncsu.edu/gis/formats.html

#

ArcCatalog Overview

The ArcCatalog application (Figure 4) is a file browser that allows a user to organize and manage various types of spatial data such as maps, datasets, metadata, services, etc. ArcCatalog includes tools to:

- Organize GIS content
- Record, view, and manage metadata.

- Search for and add content to ArcGIS applications
- Preview GIS datasets.
- Access Geoprocessing toolboxes, models, and Python scripts
- Copy, paste, delete, export spatial data

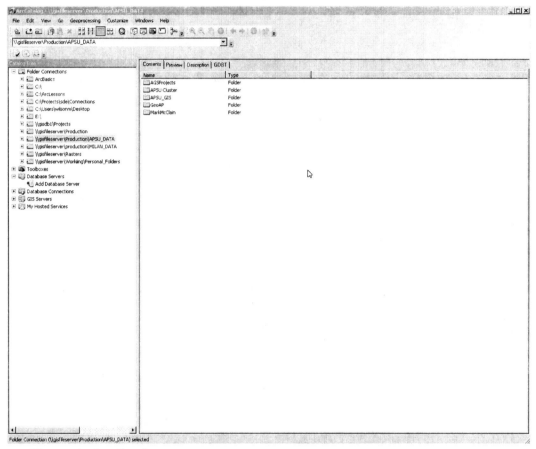

Figure 4: ArcCatalog

Before the advent of ArcCatalog, manipulating geospatial data was complicated. As noted previously, many GIS data files types are comprised of multiple files. When users attempted to copy, move, and rename these files, it was not uncommon for data to become corrupted. Using the file management functions of ArcCatalog eliminated these issues and simplified how users can move, copy, or deletes all related files with one operation rather than many.

The following exercise demonstrates how to use ArcCatalog to:

- Launch ArcCatalog
- Connect to a folder or directory
- Create a new shapefile
- Preview spatial data

- View metadata
- Launch ArcMap and open the ArcToolbox window

Launch ArcCatalog

 ArcCatalog

Figure 11: ArcCatalog Icon

1. Select Start > All Programs > ArcGIS > ArcCatalog 10.1 or 10.2 (Figure 11).
2. Similar to Windows Explorer, ArcCatalog allows GIS users to manipulate spatial data. As users move around the interface, each icon has a popup label describing the button's functionality. In some cases, users can also select "F1" to bring up more info.
3. The pane on the left side of the application contains the "Catalog Tree" (Figure 12), which gives the user access to various folders, files, database, and server connections. These folders and connections are those that the user had previously established. A folder and connection will only be available once it is established.

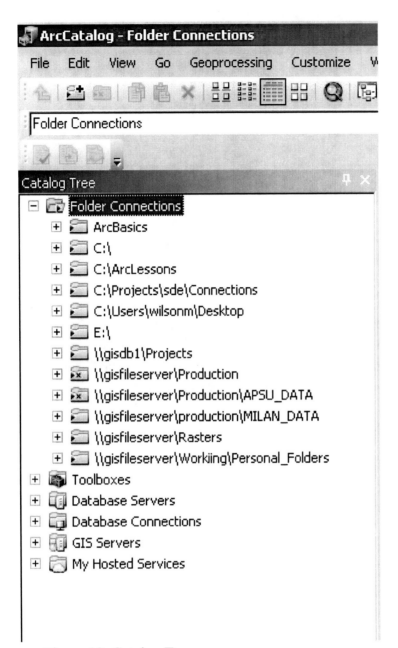

Figure 12: Catalog Tree

4. In versions of ArcGIS greater that v9, aspects of ArcCatalog are now directly integrated into ArcMap.

#

Create a New Shapefile

ArcCatalog has functionality to create various feature datasets. In the following example, we will create a new shapefile.

Make sure the "...\ArcBasics\Intro\data" folder is selected or highlighted.

1. Click on File in the Menu Bar and choose New > Shapefile....

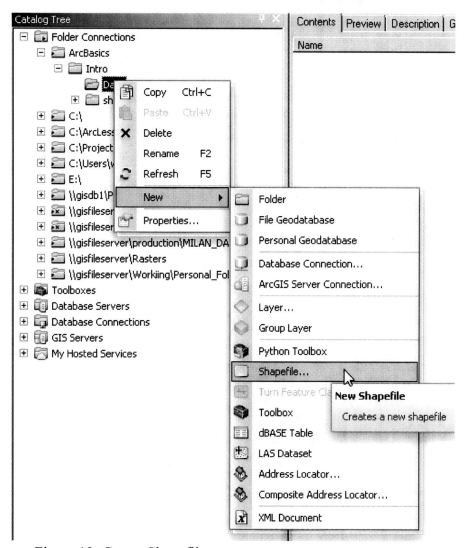

Figure 13: Create Shapefile

2. The Create New Shapefile dialog window opens (Figure 13). Name the new shapefile "arc_example" and choose the Polygon Feature Type from the dropdown menu.

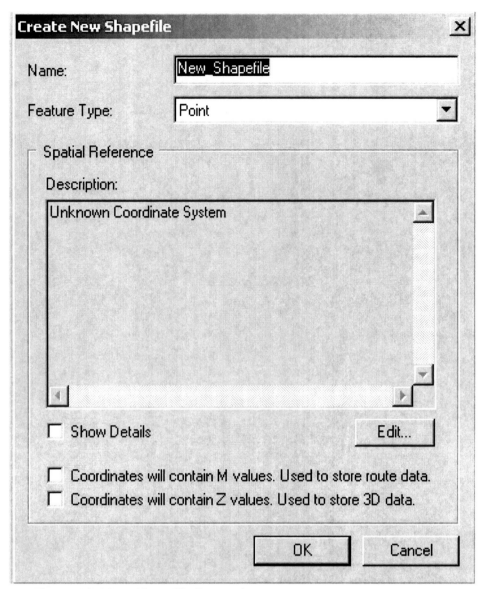

Figure 14: New Shapefile Properties

3. Spatial Reference describes where features are located in the real world. As you create the new shapefile, he or she can define the spatial reference. Click the Edit... button in the Create New Shapefile dialog then click the Select... button in the Spatial Reference Properties window (Figure 10).

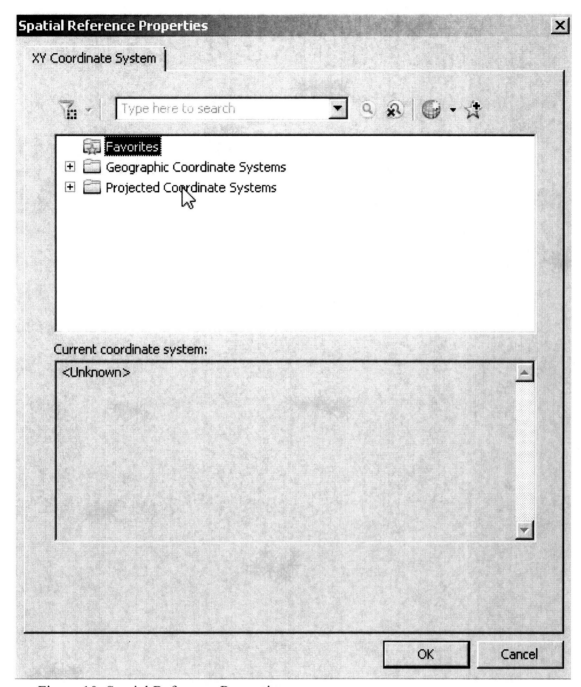

Figure 10: Spatial Reference Properties

4. Double-click on the Geographic Coordinate Systems folder and then on the North America folder.
5. Double-click on the NAD 1983.prj file.
6. Select the file and click Add. The Coordinate system is now displayed in the Spatial Reference Properties window.

7. Click OK and view the Spatial Reference Description in the Create New Shapefile dialog (Figure 10).
8. Click OK and observer the new shapefile in the ...\ArcBasics\Intro\data folder.

#

Preview an Existing Shapefile

1. In the Catalog Tree on the left side of the ArcCatalog window, navigate to ...\ArcBasics\Intro\shp\. Click on the apsu_buildings.shp shapefile. In the "Contents Widow" select the Contents tab and you will see a polygon icon, indicating that this shapefile contains polygon features.

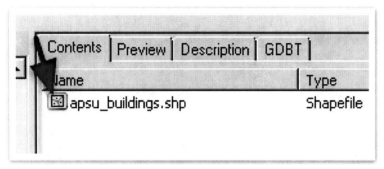

Figure 15: Polygon Icon

2. Select the Preview tab (Figure 16) to preview the shapefile
3. At the bottom of the Preview area, click on the dropdown arrow to the right of the small window labeled Preview: and choose Table. This allows you to view the shapefile's attribute table.

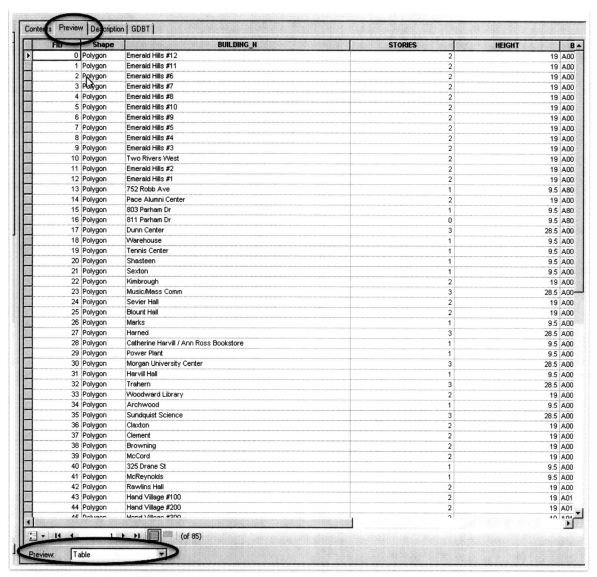

Contents	Preview	Description	GDBT

FID	Shape	BUILDING_N	STORIES	HEIGHT	B
0	Polygon	Emerald Hills #12	2	19	A00
1	Polygon	Emerald Hills #11	2	19	A00
2	Polygon	Emerald Hills #6	2	19	A00
3	Polygon	Emerald Hills #7	2	19	A00
4	Polygon	Emerald Hills #8	2	19	A00
5	Polygon	Emerald Hills #10	2	19	A00
6	Polygon	Emerald Hills #9	2	19	A00
7	Polygon	Emerald Hills #5	2	19	A00
8	Polygon	Emerald Hills #4	2	19	A00
9	Polygon	Emerald Hills #3	2	19	A00
10	Polygon	Two Rivers West	2	19	A00
11	Polygon	Emerald Hills #2	2	19	A00
12	Polygon	Emerald Hills #1	2	19	A00
13	Polygon	752 Robb Ave	1	9.5	A80
14	Polygon	Pace Alumni Center	2	19	A00
15	Polygon	803 Parham Dr	1	9.5	A80
16	Polygon	811 Parham Dr	0	9.5	A80
17	Polygon	Dunn Center	3	28.5	A00
18	Polygon	Warehouse	1	9.5	A00
19	Polygon	Tennis Center	1	9.5	A00
20	Polygon	Shasteen	1	9.5	A00
21	Polygon	Sexton	1	9.5	A00
22	Polygon	Kimbrough	2	19	A00
23	Polygon	Music/Mass Comm	3	28.5	A00
24	Polygon	Sevier Hall	2	19	A00
25	Polygon	Blount Hall	2	19	A00
26	Polygon	Marks	1	9.5	A00
27	Polygon	Harned	3	28.5	A00
28	Polygon	Catherine Harvill / Ann Ross Bookstore	1	9.5	A00
29	Polygon	Power Plant	1	9.5	A00
30	Polygon	Morgan University Center	3	28.5	A00
31	Polygon	Harvill Hall	1	9.5	A00
32	Polygon	Trahern	3	28.5	A00
33	Polygon	Woodward Library	2	19	A00
34	Polygon	Archwood	1	9.5	A00
35	Polygon	Sundquist Science	3	28.5	A00
36	Polygon	Claxton	2	19	A00
37	Polygon	Clement	2	19	A00
38	Polygon	Browning	2	19	A00
39	Polygon	McCord	2	19	A00
40	Polygon	325 Drane St	1	9.5	A00
41	Polygon	McReynolds	1	9.5	A00
42	Polygon	Rawlins Hall	2	19	A00
43	Polygon	Hand Village #100	2	19	A01
44	Polygon	Hand Village #200	2	19	A01
45	Polygon	Hand Village #300	2	19	A01

|◄ ◄ 1 ► ►| (of 85)

Preview: Table

Figure 16: Shapefile Preview

\#

View Metadata

The next tab is used to view Metadata (Figure 17).

1. Select the Description tab. The metadata for the apsu_buildings.shp file is now displayed. The Metadata tab may now be used.

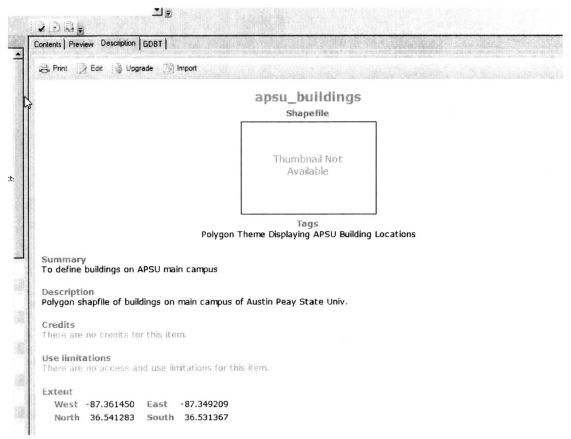

Figure 17: Metadata Toolbar

2. Hold the pointer over each Metadata button to see its function.
3. To change the Metadata data style, go to Customize => ArcCatalog Options => click on Metadata Tab. Under Metadata Style, select the style of your choosing on the dropdown arrow (Figure 18).

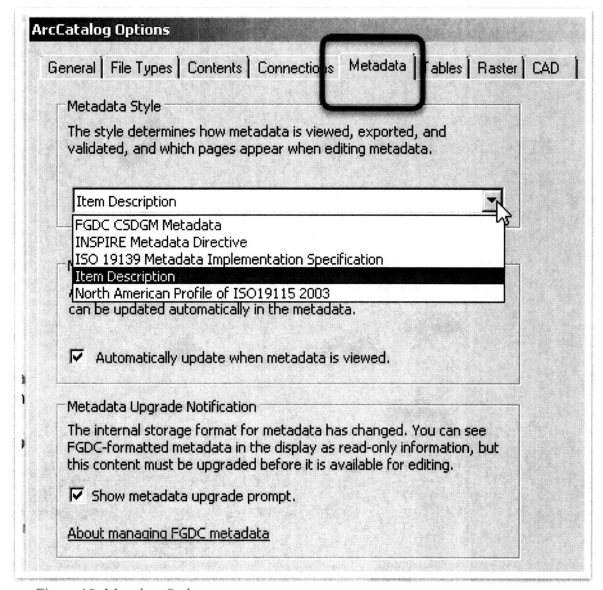

Figure 18: Metadata Style

#

ArcMap Overview

ArcMap is the main component of ArcGIS. It is used to view, edit, create, and analyze geospatial data. ArcMap also allows the user to view, symbolize and create maps. ArcMap documents are saved with the MXD extension, as the term MXD is used to designate an ArcMap document.

#

Launch ArcMap

ArcMap can be launched in three ways:
- From the Start Menu
- From ArcCatalog
- From an existing ArcMap Document (or MXD)

1. Launching from the Start Menu:

Users can launch ArcMap by clicking Start > Programs > ArcGIS > ArcMap 10.1 or 10.2. When prompted, select a new Blank Map (Figure 19).

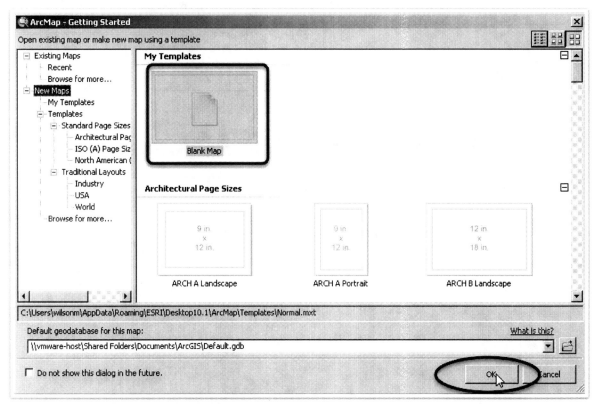

Figure 19: ArcMap Start

2. Launching from ArcCatalog:
    To launch ArcMap from ArcCatalog, click on the ArcMap icon on the standard toolbar. ArcMap will take a moment to open. When prompted, select a new empty map.

3. Launching from an Existing MXD:
    Navigate to a previously saved MXD and double-click on it. The MXD will open to the previously save map.

#

Add and Remove Data Layers

To start getting familiar with the ArcMap interface, add both a shapefile and image (MrSID) to a new, empty ArcMap document.

1. Launch ArcMap from the Start Menu: Click Start > Programs > ArcGIS > ArcMap 10.1 or 10.2. When prompted, select a new Blank Map.

Figure 20: The Standard Toolbar

2. If necessary, turn on the Standard Toolbar (Figure 20) by clicking on the Customize menu item > Toolbars > Standard (Figure 21).

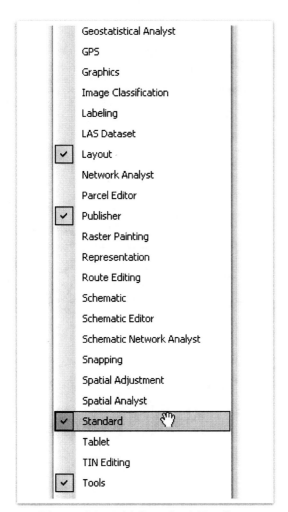

Figure 21: Add Standard Toolbar

3. Click the ⬩ (Add Data button) on the Standard Toolbar. In the Add Data dialog, navigate to ...\ArcBasics\Intro\shp folder, highlight apsu_buildings.shp, and click the Add button. Notice that the data layer, apsu_buildings, is added to the "Layers" Data Frame in the ArcMap Table of Contents.

4. Add a Digital Orthophotography Quarter Quad (DOQQ) to ArcMap. A DOQQ is an ortho photo that represents 1/4 of a USGS topo map. Click the Add Data button on the standard toolbar. In the Add Data dialog (Figure 22), navigate to "...\ArcBasics\Images" folder, and open the "Rasters.gdb" File Geodatabase add the "APSUCampus" image to your map. Click "OK".

    * If you receive a "Spatial Reference" warning, ignore it. In later sections, you will learn about identifying and setting a spatial reference and coordinate systems. *

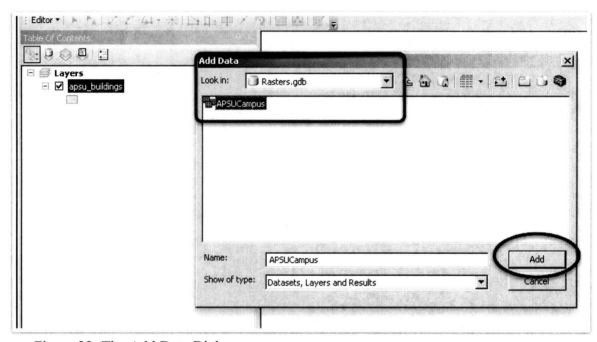

Figure 22: The Add Data Dialog

5. To remove a data layer, right-click on the name of the layer in the Table of Contents window and select Remove from the context menu that appears (Figure 23).

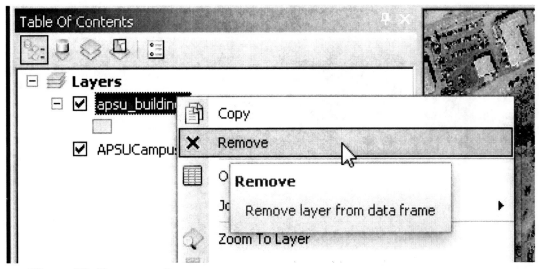

Figure 23: Remove a Layer

#

As we have seen previously, one can easily add GIS data to a map. In this exercise we will add some ESRI Basemap data to a map. The "ESRI Basemaps" are reference layers that are made freely available from ESRI. In order to utilize these layers, one must be connected to the Internet.

1. Open a blank MXD.

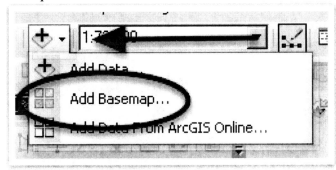

Figure 24: Add Basemap

2. Click on the drop-down arrow next to "Add Data" Button (Figure 24).
3. Browse the available basemaps and select a map to add to the map (Figure 25).

Figure 25: Available Basemap

4. To add a map you can either double-click it or you can select it and click "Add".

#

According to ArcGIS Resources[6], ArcGIS Online is a:

Collaborative, cloud-based platform that lets members of an organization create, share, and access maps, applications, and data, including authoritative basemaps published by ESRI. Through ArcGIS Online, you get access to ESRI's secure cloud, where you can manage, create, store, and access hosted web services, and because ArcGIS Online is an integral part of the ArcGIS system, you can use it to extend the capabilities of ArcGIS for Desktop, ArcGIS for Server, ArcGIS applications, and ArcGIS APIs and Runtime SDKs.

In ArcGIS Desktop you can easily add data from ArcGIS Online via the following steps:

1. Open a blank MXD.

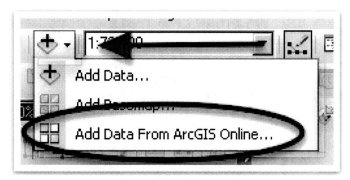

Figure 26: Add ArcGIS Online Data

2. Click on the drop-down arrow next to "Add Data" Button (Figure 26).
3. Browse the available ArcGIS Online data and select a map to add (Figure 27).

Figure 27: Browse ArcGIS Online

4. To add a map you can double-click it to add it to the MXD.

#

The MXD

The MXD is a container for all the spatial data being used in ArcMap. An MXD is a collection of data layers and cartographic layouts. MXD store references to the locations of the data sources (e.g., shapefiles, coverages, image files, etc.) **NOT** the data themselves.

When attempting to send GIS info via email or other methods, it is common for those new to GIS to transmit only the MXD file. To transfer data, the user must send the feature datasets (e.g., shapefiles, coverages, image files, etc.) that are referenced in the MXD. The MXD stores only references!

Save your untitled map document, as follows:

1. Click "File" in the menu bar and select "Save As" (or Save) from the dropdown list.
2. In the "Save As" dialog, navigate to the folder in which you want to save your map document (e.g. ...\ArcBasics\Intro).
3. Type in a file name such as "APSU", and click the "Save" button. Notice that the file name "APSU" now appears in the ArcMap title bar.

The ArcMap Interface (Figure 28)

Key components of the ArcMap interface are:
- The title bar (1), menu bar (2), and toolbars (3);
- Two side-by-side windows: the Table of Contents (4) and the map display window (5);
- The status bar (6).

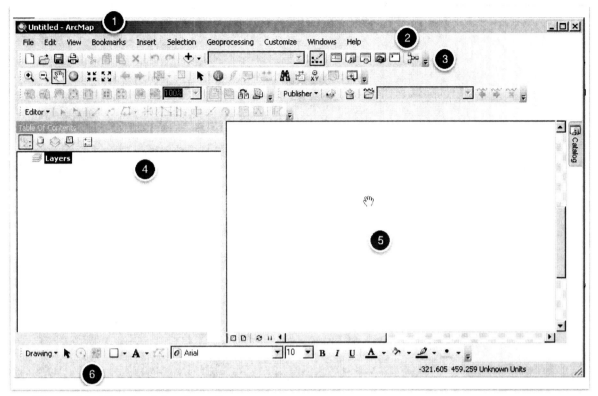

Figure 28: The ArcMap Interface

The Title Bar (1) at the top of the ArcMap window displays the name of the map document.

The Menu Bar (2), located just below the Title Bar contains a series of menu items: File, Edit, View, Bookmarks, Insert, Selection, Geoprocessing, Customize, Window, and Help.

As with any windows menu item, clicking on an item opens a dropdown menu with numerous options. Hover the mouse over an item to see a popup explanation of its function. To select an option on the dropdown menu, click and release the mouse button on the option.

ArcMap includes a variety of Toolbars. Click on the Customize menu item and select Toolbars to see all of the toolbars that are available to you. A check mark next to the toolbar name indicates that it is visible. For now, ensure that the Standard Toolbar and the Tools Toolbar (Figure 29) are turned on, as shown below:

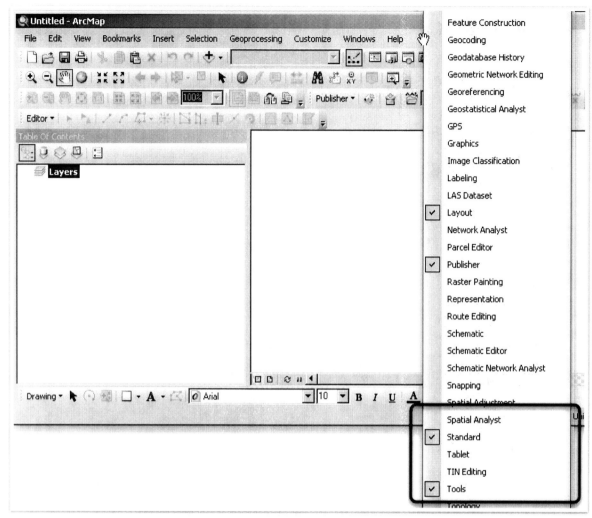

Figure 29: The Standard and Tools Toolbars

The Standard Toolbar (Figure 30) contains buttons that give you fast and easy access to many of the Menu Bar options. Hover over a button to see a popup description.

Figure 30: The Standard Toolbar

The Tools Toolbar includes tools that allow you to interact with the data displayed in the map display window to carry out specific tasks. Typically, the pointer changes when a tool is used.

Figure 31: The Tools Toolbar

ArcMap provides a number of other toolbars, such as:

Draw - for drawing graphics and adding text
Editor – for editing spatial data
Layout – for working with layouts
Effects – for altering the display of spatial data

Take a look at these and some of the other toolbars and their tools.

Figure 32: Handle to move the Toolbar

Most of the toolbars can be docked or floated in ArcMap by clicking and dragging the toolbar handle (see Figure 32). You may set the toolbars to suit your workflow.

You can access the toolbars list without using the Customize menu by right-clicking any toolbar or the status bar. To quickly hide or turn off a toolbar, click its "Close" button.

The Status Bar (Figure 33), located at the bottom of the ArcMap window displays the coordinate position of the mouse pointer in the display window and information about the progress of several operations, e.g., printing and exporting.

Figure 33: The Status Bar

The Table of Contents (TOC) (Figure 34), located on the left side of the ArcMap window, lists all of the layers (and the layer type) that you have added to your map and shows the symbols that are used to represent the features in each data layer.

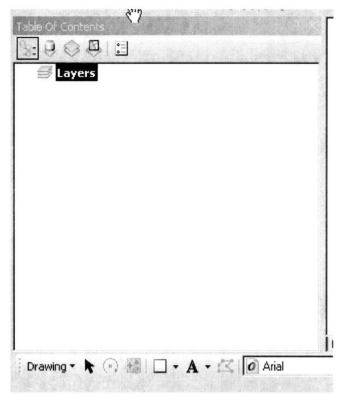

Figure 34: The Table of Contents (TOC)

The check box (Figure 35) next to each layer name indicates whether it is currently visible (the layer is "On") in the map display window to the right of the TOC.

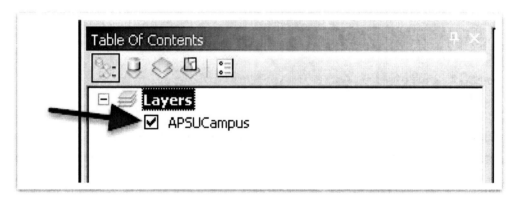

Figure 35: Layer Toggle

The layers are drawn in the order they are listed in the TOC. The layers at the bottom of the list are drawn underneath those listed at the top.

At the top of the TOC, there are several buttons (Figure 36): a "List by Drawing Order" button, a "List by Source" button, a "List by Visibility" button, "List by Selection" button and an "Options" button.

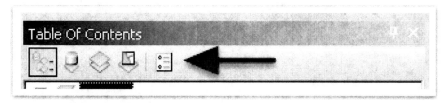

Figure 36: TOC Buttons

When you select the "List by Drawing Order" tab, you see the drawing order of the layers and you can move a layer up or down in the TOC by dragging and dropping it.

1. Move the apsu_buildings layer to the bottom of the TOC and notice what happens.
2. Now, move it back to the top of the TOC.

When you select the "List by Source" tab, layers are sorted by where they are stored on disk/network/connection. This is useful during editing when you want to edit all layers in a given folder or database. You cannot change the drawing order of layers when the Source tab is selected.

The "List by Selection" tab allows you to choose the layers from which features can be selected. You will learn more about selecting features later in this book.

Within the TOC, data layers that you add to your map are grouped into one or more data frames. A data frame is a group of data layers that you want to display together. When you create a map (or map document) in ArcMap, it always contains at least one data frame, with the default name of "Layers" that is listed at the top of the TOC. You can change the data frame name to something more meaningful. You will see the advantages of utilizing multiple data frames in the "Making a Layout" Section.

In ArcMap, geographic information is displayed on a map as layers. Each data layer represents a specific type of feature such as rivers, lakes, archaeological sites, political boundaries, etc. A data layer does not store the actual geographic data; instead, it references the data contained in coverages, shapefiles, geodatabases, images, grids, and so on. Referencing data in this way allows the layers on a map to automatically reflect the most up-to-date information in your GIS database.

Data View and Layout View
ArcMap provides two different views: data view and layout view (See Figure 37 and Figure 38, respectively). Use data view when you want to browse, edit, and/or analyze the geographic data on your map. Layout view is used to prepare finished maps for printing and publication.

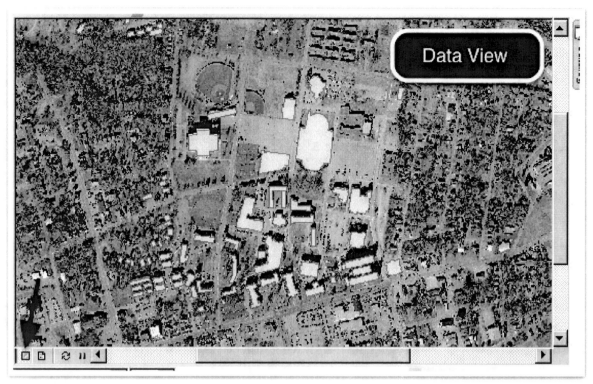

Figure 37: Data View

Figure 38: Layout View

To switch between data and layout views:

1. Click the "View" menu item and select either "Data View" or "Layout View" from the context menu that appears.
2. Alternatively, you can use the "Data View" and "Layout View" buttons located in the lower left portion of the view window to switch between these two views.
3. When you switch to Layout View, the Layout toolbar (Figure 39) is automatically added to the ArcMap window. These tools allow you to zoom in and out, pan, and zoom to set extents (e.g. full page, 1:1) on the layout.

    The "extents" is the area covered by a layer or map.

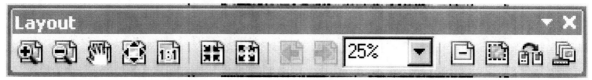

Figure 39: Layout Toolbar

#

Data Frame Properties: Map Units

Map units are the horizontal measures (feet, meters, etc.) in which distances are calculated in a data frame. They are different from the Display units, which are the horizontal measures in which distances are actually displayed on the screen. For example, you might be using data stored in State Plane feet (so the map units are set to feet), but making distance measurements in meters (the display units).

Map units are set when a coordinate system is selected for a data frame. The data frame coordinate system is automatically set to the coordinates of the first data layer added to the data frame. This feature is very useful when data stored in different coordinate systems are going to be used together. For example, you might want to display a data layer with UTM coordinates on top of a layer stored in latitude/longitude. ArcGIS reprojects the data on the fly to whatever coordinate system is specified for the data frame. The user in this situation cannot alter the map units.

However, on-the-fly projection does not work correctly unless all the data layers in a data frame have their coordinate systems defined - usually in a projection file with a .prj extension. If you add a layer that does not have a defined coordinate system, ArcMap displays a warning message: "One or more layers is missing spatial reference information. Data from those layers cannot be projected." (On-the-fly projection will be covered in more detail in the Section on "Coordinate system and Projection".)

When the term "on-the-fly projection" is used, it means that a layer is dynamically projected into another coordinate system. For

example, if there are 2 layers in a map, a layer in decimal degrees and a map in State Plane coordinates, ArcMap will automatically convert the State plane layer to decimal degrees. This change is not permanent.

Finally, if a data frame does not have a selected coordinate system (probably because the first added data layer didn't have a .prj file), then the map units should be set by the user to whatever coordinate system is appropriate for the data in the data frame. In this case, all data layers should be stored in the same coordinates; otherwise, serious misalignment issues can arise.

Your ArcMap map document, apsu.mxd, currently contains one data frame, named Layers, and two data layers. The first data layer that you added, apsu_buildings, is a shapefile that includes a projection file (apsu_buildings.prj). Because the apsu_buildings shapefile includes a projection file, when you added it to your map document, ArcMap set the data frame's coordinate system and map units based on information in the projection file.

1. Right-click on the data frame (named Layers) and select "Properties" from the context menu that appears (Figure 40).

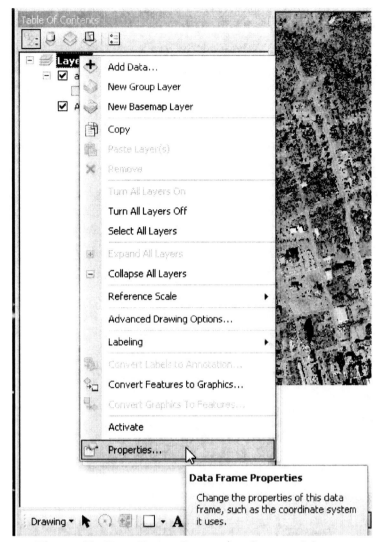

Figure 40: Data Frame Properties Context Manu

2. Click on the "General" tab in the "Data Frame Properties" dialog (Figure 41) and notice that the map units are set to feet (this control is disabled).

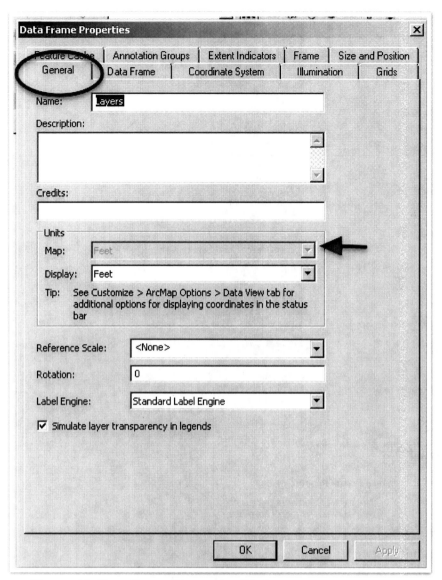

Figure 41: Data Frame Properties Dialog

3. Click on the "Coordinate System" tab (Figure 42) and note that the coordinate system has also been set. Cancel out of the "Data Frame Properties" dialog when you have finished examining its contents.

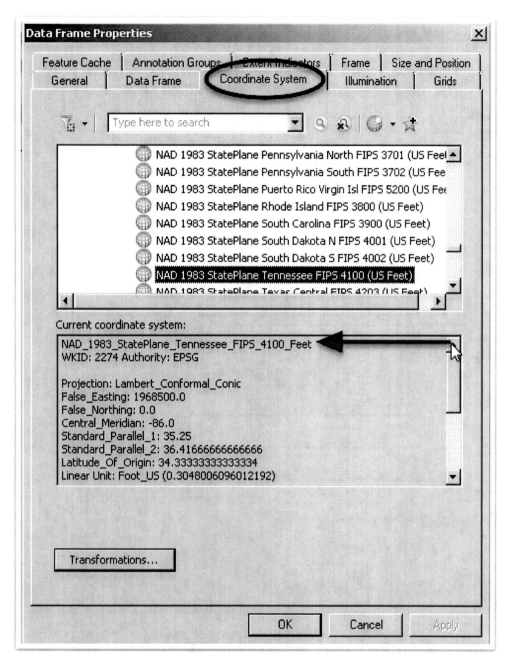

Figure 42: Data Frame Coordinate System

#

Data Frame Properties: Display Units

Display units (Figure 43) are used to report measurements you make using the measure tool, dimensions of shapes, distance tolerances, and offsets. Using the Data Frame Properties dialog, you may set display units to any unit of measure that is convenient or necessary for a specific task. Display units are independent of map units.

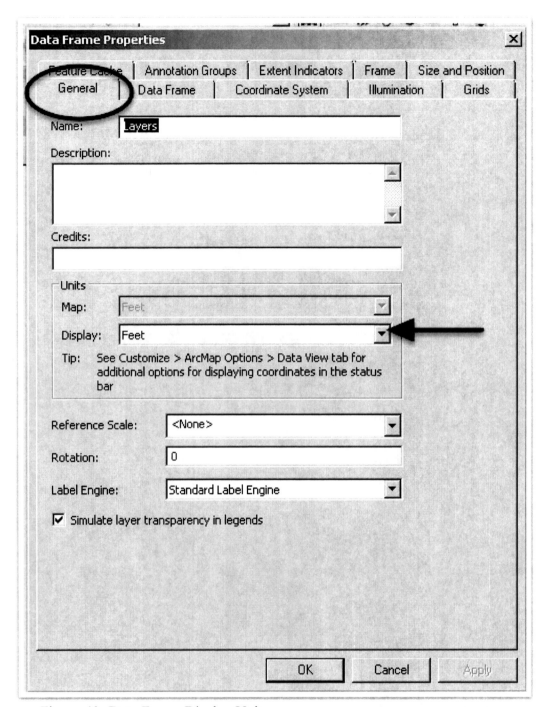

Figure 43: Data Frame Display Units

Note: Map and Display units will be covered in more detail in the "Displaying Spatial Data" Section.

#

ArcToolbox

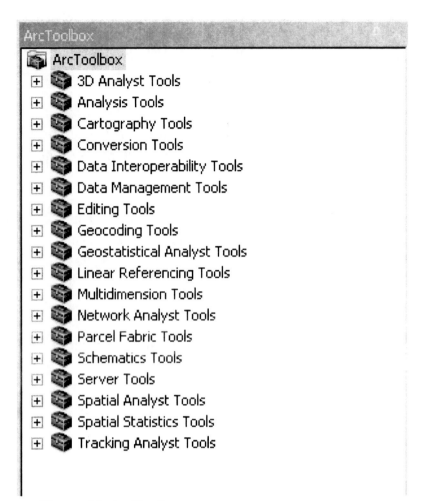

Figure 44: ArcToolbox

The various toolboxes in ArcToolbox (Figure 44) are references or shortcuts to ArcGIS tools. They are similar Windows shortcut or layers within ArcMap. If you delete a tool from inside a toolbox the tool is permanently deleted. The ArcToolbox may be reached by clicking on the ![icon] (ArcToolbox icon) on the Standard Toolbar. The main toolboxes to be examined are:

- Analysis Tools – for performing geoprocessing functions (also contains the most commonly-used analytical tools, which will be introduced in later sections).
- Conversion Tools – allow the import and export of geospatial data and other data formats compatible with ArcGIS.
- Data Management Tools – used to develop, manage, and maintain feature classes, datasets, layers and raster data structures.

Toolboxes contain toolsets and tools. Toolsets are used to group collections of tools together into logical groupings. A tool is an entity that performs a specific geoprocessing task such as generalizing lines. Clicking on a tool will display a tool description in the Status Bar. Double-Click to open the tool. There are three types of tools, system tools, models, and scripts.

- System tools let you analyze and modify spatial data.
- Models run a chain of tools in sequence.
- Scripts take advantage of the command line to run tools in sequence and are useful for batch processing. For example, converting many datasets to another format or running the same model with a series of different input datasets.

Use the ArcToolbox icon in ArcCatalog or ArcMap to open ArcToolbox. This will add a new frame to the main ArcCatalog or ArcMap interface. Navigate through the ArcToolbox toolboxes to familiarize yourself with the tools available. When you are finished, close the ArcToolbox Window.

#

ArcGIS Desktop Help

The ArcGIS Desktop Help system contains a variety of information. Help can be accessed in a variety of ways.

1. Clicking on the Help menu item gives you access to the complete ArcGIS Desktop Help system.

The ArcGIS Desktop Help window (Figure 45) has three tabs: Contents, Search, and Favorites. The help documents are organized by topic and often by application that is ArcCatalog, ArcMap, etc.

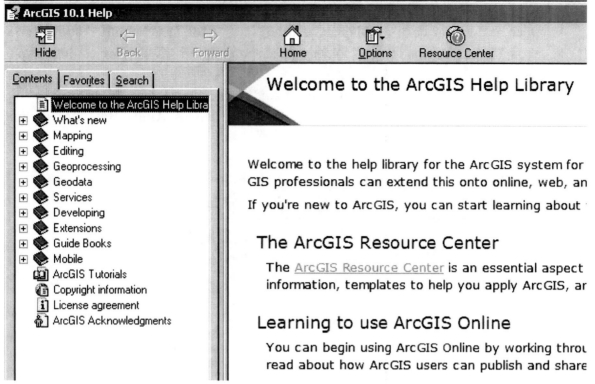

Figure 45: ArcGIS 10.1 or 10.2 Help

2. You can also press the F1 key to access the help window.

#

Intro Exercise 1

Because your employer is new to GIS, you are charged with reviewing various GIS datasets. You will need to load imagery and inspect the layers for accuracy. Does the data match the aerial photos?

In this exercise, you will load various datasets and evaluate how well each layer fits the aerial photos. The data needed is located in the "...\ArcBasics\ Intro\shp" and "...\ArcBasics\ Images" folders.

The main objective of this exercise is to explore the capabilities of ArcMap.
Data Required:

- APSUCampus
- Apsu_Buildings
- Parking
- Sidewalks

1. Open a new instance of ArcMap and use the 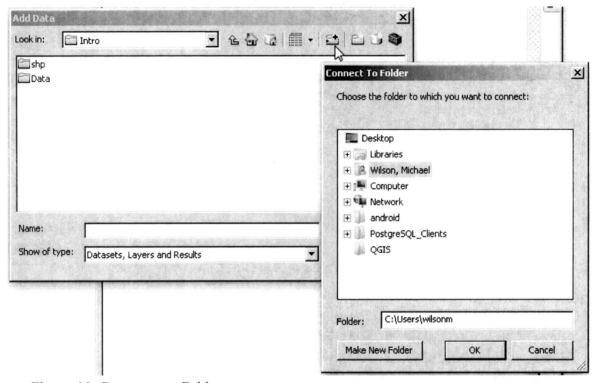 (Add Data button) to add the aerial photo ("...\ArcBasics\ Images" folder) and shapefile layers ("...\ArcBasics\ Intro\shp" folder).

2. Use the "Connect to Folder" button in the Add Data dialog (Figure 46) to create a connection to the ArcBasics folder. Add the APSUCampus, APSU_Buildings, Parking, and Sidewalks layers.

Figure 46: Connect to a Folder

3. Make sure that the "APSUCampus" raster is located at the bottom of the TOC.
4. See if you can figure out how to change the colors of the various shapes.
5. Use the navigation tools to move around the map. Right-Click on the "APSU_Buildings" layer and click to "Zoom to Layer".
6. How well does the "Sidewalks" layer match up with the APSUCampus Raster? Inspect the other layers for a match.

Chapter 2 - The GIS Workspace

Introduction

When opening ArcMap or ArcCatalog for the first time. It is often the case where menus are hanging in space or not exactly where expected. In this brief section, the user will learn how to customize the ArcGIS workspace. As previously seen, the ArcGIS workspace is comprised of three components. In this section one will be introduced to setting up ArcMap and ArcCatalog.

#

Real World Application (RWA)

Before beginning a new project, it is always time well spent to optimize and arrange your workspace. I always think of a surgeon in the midst of an operation. Time is spent setting up the surgical instruments so that the surgeon can quickly pick up tools without spending much time searching. As a GIS professional, the more time you spend actually doing your project (and the less searching for tools and data) the better. Time spent on preparation is time well spent.

#

GIS Workspace Objectives

1. Setting up Folder Connections
2. Manipulating ArcGIS Menus and Toolbars
3. ArcCatalog in ArcMap
4. Search Functions
5. Setting up Geoprocessing history.
6. Backing up your Data

#

Connect to a Folder

In ArcCatalog (Figure 47), folder connections enable the user to access your file-based data (such as coverages, shapefiles, images, etc.). All of the folder connections that you make are listed in the Catalog tree (Figure 12). Once connected to a folder, one can browse its contents in the Catalog, including the contents of any of its subfolders. Connections created in ArcCatalog are available in ArcMap. The connections are also persistent across sessions.

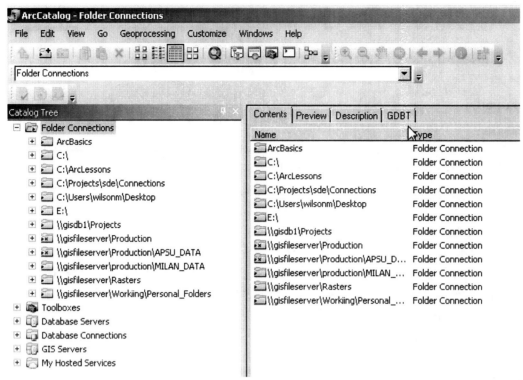

Figure 47: ArcCatalog

One can connect to any folder on your hard drive or network. Although a user can connect to a top-level folder, it is advisable to connect to subfolders. Connecting to subfolders increases speed and efficiency.

The user should think of folder connections as "bookmarks" Make a separate connection to each of your most frequently used GIS folders. Using connections will give the user easy access to data. When setting up your workspace, it is important to spend time creating any connections needed.

Use the following steps to make a folder connection:

1.   Click on the Connect to Folder icon in the menu bar.

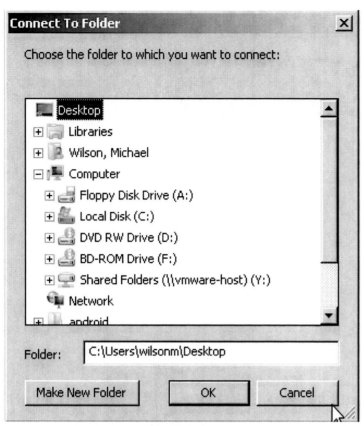

Figure 48: Connect to a Folder

2. Navigate to the folder where the data for Chapter 1 of this course are saved (e.g., C:\ArcBasics\) in the Connect to Folder dialog.

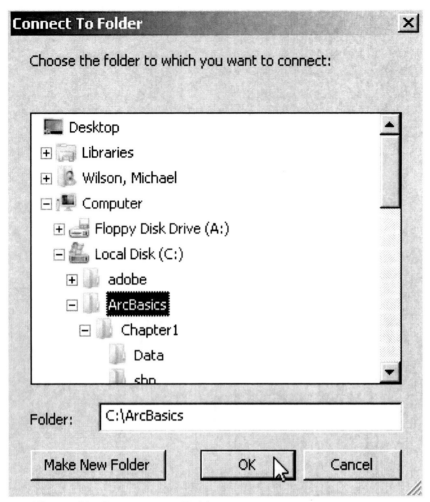

Figure 49: Connect to a Folder

3. Click "OK". The folder now appears in the Catalog Tree (Figure 49).

To remove a folder connection, right-click on the folder path\name in the Catalog tree and select "Disconnect Folder" from the context menu that appears.

#

Manipulating ArcGIS Menus and Toolbars

To start getting familiar with the ArcMap interface, let's add both a shapefile and image (MrSID) to a new, empty ArcMap document.

1. Launch ArcMap from the Start Menu: Click "Start" > "Programs" >" ArcGIS" > ArcMap 10.1 or 10.2. When prompted, select a new "Blank Map".

Figure 20: The Standard Toolbar

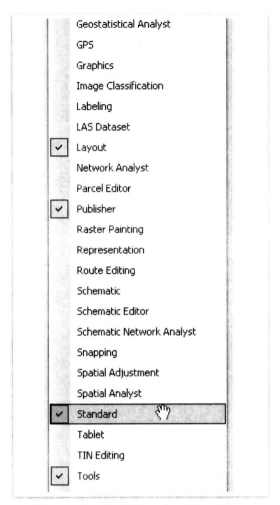

Figure 21: Add Standard Toolbar

2. If necessary, turn on the "Standard Toolbar" (Figure 20) by clicking on the "Customize" menu item > Toolbars > Standard (Figure 21). Notice that there are many other tools available. The tools that are most likely to be used include:

- Standard (Figure 20)

Figure 20: The Standard Toolbar

- Tools (Figure 50)

Figure 50: The Tools Toolbar

- Layout (Figure 51)

Figure 51: The Layout Toolbar

- Editor (Figure 52)

Figure 52: The Editor Toolbar

Take a moment to arrange these tools in ArcMap. By arranging the toolbars, the user will be able to learn their position and be able to easily find tools when needed. Developing muscle memory for these tools will move the novice user one step closer to GIS mastery.

#

ArcCatalog in ArcMap

With recent versions of ArcMap, you can access ArcCatalog directly from ArcMap. ArcCatalog can be accessed via a tab on the right hand side of the ArcMap interface.

1. Click the ArcCatalog tab to open the catalog window (Figure 53).

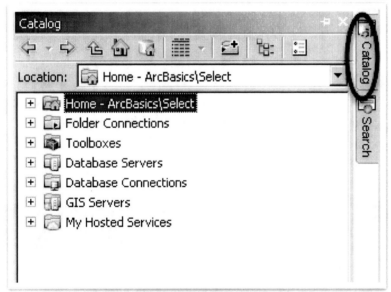

Figure 53: ArcCatalog Window

2. Notice the functions at the top of the "ArcCatalog" window. Nearly all ArcCatalog functionality is available.

Use the following steps to add the "ArcCatalog" tab to the ArcMap window when it is not visible.

1.   Click on "Windows" in the menu bar.
2.   Click on "ArcCatalog" (Figure 54).

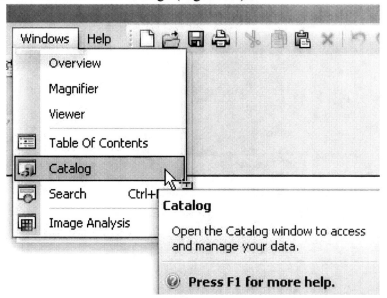

Figure 54: Windows Menu

3.   The "ArcCatalog" tab will appear in the interface.
#

Search Functions

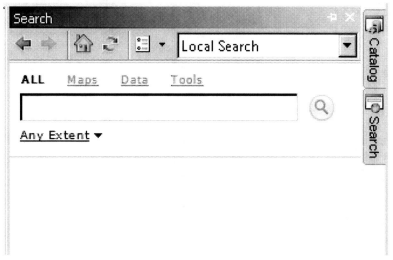

Figure 55: Search Window

ArcMap has a universal search tool built into the interface. The search window can be accessed via the "Search" tab located on the right side of the ArcMap interface (Figure 55). In cases where the Search tab is not visible, click on "Windows" in the menu bar and click on "Search" (Figure 56).

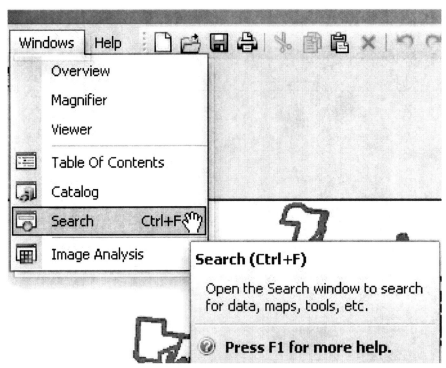

Figure 56: Windows Menu - Search

The search widow allows the user to complete a variety of searches. These searches include: universal (All), Map, Data, and Tool searches. In order to search for maps and data, the search options need to be configured:

1.   Open the Search Tab.
2.   Click on the "Search" options button (Figure 57).

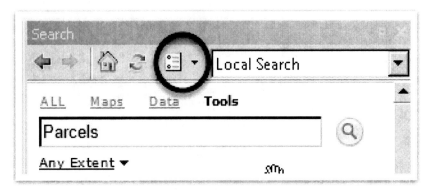

Figure 57: Search Options Button

3.   The Search Options dialog will open (Figure 58).

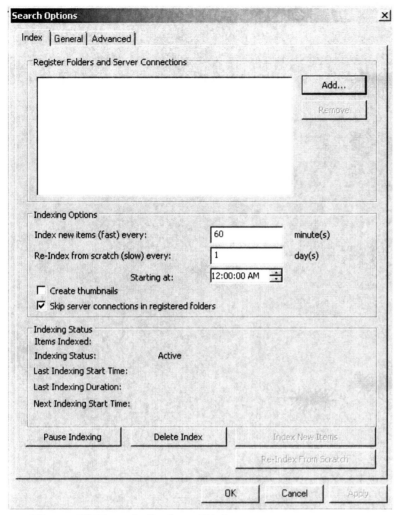

Figure 58: Search Options Dialog

4. Once in the options, you can add folders to be indexed. Index folders will automatically be searched when using the All, Map, and Data searches.
5. To add a folder, click the "Add" button and navigate to the folder you would like to add to the search index.

The Tools search will locate various tools that match the search criteria. You will learn more about the tools in the "Geoprocessing" section.

#

Setting up Geoprocessing History

In later sections, Geoprocessing will be discussed. Each time the user runs a geoprocessing function; it saves information to a log file. By default, ArcMap stores logs going back two weeks. Storing so much data in ArcMap has a tendency to make it slower. In this exercise, one will learn how to adjust the geoprocessing log.

1. Launch ArcMap from the Start Menu: Click" Start" > "Programs" > "ArcGIS" > ArcMap 10.1 or 10.2. When prompted, select a new "Blank Map".

2. Click on the "Geoprocessing" menu item and then click "Geoprocessing Options..." (Figure 59).

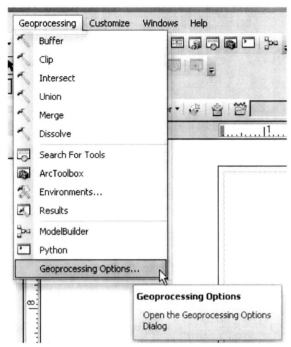

Figure 59: The Geoprocessing Menu

3. The "Geoprocessing Options" dialog will open. In the "Results Management" section, change the "Keep the results younger than" drop-down to "3 Days" (Figure 60).

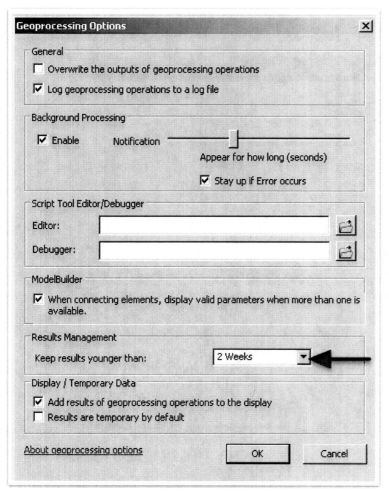

Figure 60: The Geoprocessing Options Dialog

4. Click "OK" to apply.

#

Backing up your Data

When working with GIS data, there maybe cases where the user would like to make a quick copy of your data before editing or making any changes. In this exercise, this section will discuss how to make a quick backup of your data.

1. Open the Workspace.mxd located in the "...\ArcBasics\Workspace" folder.
2. Right-click on "Parcels" layer in the TOC. From the context menu, click on "Data" and then "Export Data".
3. In the "Export data" dialog make sure the "Export" drop-down list shows "All Features".
4. Change the "Output Feature Class:" to "..\Workspace\Data" folder, name the layer ParcelsCopy.shp.
5. Click "OK". When prompted, click yes to add the new layer to the TOC (Figure 61).

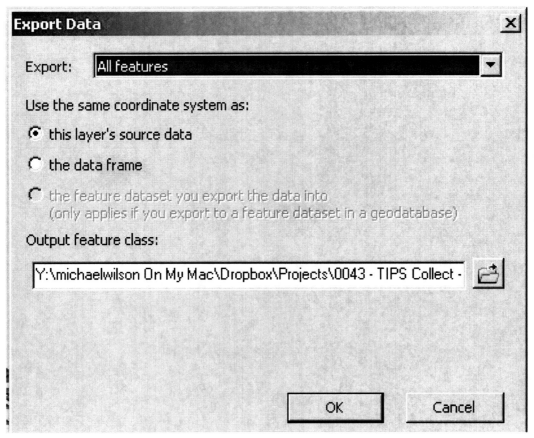

Figure 62: Export Data

6. There now should be two layers in the TOC, "Parcels" and "ParcelsCopy". Take a moment to explore each data set.

Chapter 3 - Project Management and GIS

Introduction

Like any new endeavor, many people immediately start work to create their first GIS map without any planning. The temptation to jump right in and begin building is always strong. From experience, it is better to spend some time initially planning out your project. Regardless of the complexity of a project, time spent before you begin building will be well spent developing a simple project plan.

#

Module Objectives

In this module, we will discuss:
* Defining a goal
* Setting a next action
* Listing action steps
* Identify assistance needed
* Plan review

#

Project Organization

By failing to prepare, you are preparing to fail.
Benjamin Franklin

Several years ago, I found an article that entitled "Easy Project Organization in 10 Minutes Flat" (http://goo.gl/VUE2a). Although planning a large project in 10 minutes is not recommended, the basic steps in the article can be used to develop a simple project plan that will give you a good foundation for success. In this chapter, I have adapted the 10-minute project plan for use on most GIS projects. Time spent planning on the front end of a project is significantly "cheaper" than starting to plan when you have already started to work. Basically, if you need to touch something again that you already completed due to lack of planning, it will cost you both time and money. Starting a project with even a couple of minutes of planning, will greatly improve your chance of success.

#

Define

The initial step of the planning process should be to define the project. By defining the project goal, you are building a foundation for success. I recommend using **SMART** goals. Goals need to be Specific, Measurable, Attainable, Relevant, and Timely.

Specific

A specific goal will answer the following questions:

What: What do I want to accomplish?

Why: Specific reasons, purpose or benefits of accomplishing the goal.

Who: Who is involved?

Where: Identify a location.

Which: Identify requirements and constraints.

Measurable

A measurable goal will usually answer questions such as:

How much?

How many?

How will I know when it is accomplished?

Attainable

An attainable goal will usually answer the question:

How: How can the goal be accomplished?

Relevant

Relevant goals (when met) drive the team, department, and organization forward. A goal that supports or is in alignment with other goals would be considered a relevant goal.

A relevant goal can answer yes to these questions:

Does this seem worthwhile?

Is this the right time?

Does this match our other efforts/needs?

Are you the right person?

Is this acceptable for correction?

Time-bound

A time-bound goal will usually answer the question:

When is this project due?

When do I need to start?

(Via SMART criteria - Wikipedia, the free encyclopedia - http://en.wikipedia.org/wiki/SMART_criteria)

By taking the time to outline the goal of the project before beginning, one will find that it may be easier to accomplish.

#

Identify

Many people struggle with getting started on a project. This is particularly true for larger projects. When looking at a project in its entirety, the task can often be daunting. The key to success is always identifying the next action that needs to be accomplished to move a project forward. The next minute should be set aside in order to identify next physical step of your project. By identifying this next step, you are setting yourself on the path to success. Instead of generalizing the next step into a large group of what needs to be done, try breaking it down into an easier action that seems more attainable. An example of this would be if you were in need for base data for your project. The next step could be as simple as creating a list of needed data layers. Although this may seem obvious, this step is often overlooked because of its simplicity.

Using this method will lay the groundwork for the rest of the project. Visualizing the end result may give a solid start and end to your project.

#

5-Minutes-Organize

After identifying the next action step, use the next five minutes to organize the next step of your project. A common pitfall is to become overly detailed when creating a list of tasks. Be sure to use common sense when making this task list. Don't hesitate to list out tasks and then worry about sorting the order. It is often easier to brainstorm the task list and then sort.

Keep in mind that the whole project does not need to be planned in five minutes, but proper organization of tasks which creates a natural flow will aide in the project moving into the next phase. Taking the time to list the first couple of steps will go a long way toward moving the project forward. Another 12-minute project planning session can always be done along the way as needed to solidify project organization.

#

2-Minutes-Assistance

The next two minutes can be used in order to identify anyone needed to help in the project. Managing a schedule, is much easier when working alone. When collaborating with others, the priorities of other are often beyond your control. By identifying who you are working with, you can identify potential bottlenecks. Making a list of these people can be beneficial in the long run.

#

Review

Once you have completed the previous steps, take a moment to review. Envision what success will look like. Once you have a picture of success in your mind, does the plan you have written out get you to this goal? Identify any problem areas that may need tweaking and deal with them before proceeding. The next course of action after doing this must truly be a next course of action. Actions must flow naturally and be void of awkward dependencies.

By spending a little time planning your project, you have started your project off on the right foot. This enables ideas to constantly flow and creates cooperation between both ideas and actions. While it may be true that a simple plan may not pan out for more intricate projects, a vast majority of projects will benefit in these 12 minutes.

Reference: http://www.lifehack.org/articles/productivity/easy-project-organization-in-10-minutes-flat.html

#

Review

No project plan is perfect. Even if you could develop the perfect plan, life does not pause just because you are working on a project. As you work on your projects, always build in time to review your work. By doing regular reviews, you can identify problem areas and make adjustments. There is absolutely no shame in needing to adjust your plan. In my experience, adjustments are common.

#

Post Project

I once read that the three goals every project should do include:

- Build profitable
- Build your strengths
- Encourage team building

When you complete a project, take some time to review. Answer the following questions, and use the answers to influence how you plan for future projects.

1. Was this project profitable? A profit could be financial benefit, the addition of knowledge and/or the learning of new skill sets.
2. Did this project play to your strengths? Projects need to help you work toward your long-term goals. Since you cannot always pick your own projects, use the planning process as a way to help build your strengths and meet your goals.
3. When you complete a project, if you or your team is experiencing burnout, there is a problem. Burnout is a common issue; it leads to low moral, decreased productivity and increased staff turnover. If, after a project, you or you team is experiencing burnout, look at your project plan and see if you can identify areas that could be changed in the future to minimize this type of problem.

Chapter 4 - Displaying Spatial Data

Introduction

In ArcGIS, vector data is represented by points, lines, and polygons that are defined by a set or sets of [X,Y] coordinates. Map data is only a small part of GIS data. The difference between GIS data and CAD data is that the spatial features in the GIS are linked to attribute tables (the Information Systems part of GIS). Because of this linked information, geographic features can be selected and manipulated based on the features attributes.

For example, if a user has parcel information with acreage as an attribute, one can classify and display the parcels based on size ranges. By adding a layer for flood zone data, the user can select all parcels that are located in the floodway.

#

Real World Application

In most cases, you are learning GIS either because you want to use GIS as part of your current job, or because you are looking to become a GIS professional. Let's take another quick look at the definition of GIS:

What is GIS?

A Geographic Information System (GIS) integrates hardware, software, and data for capturing, managing, analyzing, and *displaying all forms of geographically referenced information*.

I have put in italics the key part of the definition for this section, *displaying geographic information*. To perform the basics of GIS, one needs to be able to visualize information. This section will outline the basics of being able to view spatial data.

#

Objectives

In "Displaying Spatial Data", you will learn how to do the following:

*   Map Scale and Zoom Tips
*   Turning Data Layers On and Off
*   Ordering Data Layers in the TOC
*   Layer Properties
*   Set Display Units and Measure Distance on the Map Display
*   Display an Attribute Table
*   Select Features Interactively

- Use the Identify Tool to See the Attributes of a Feature
- Map Tips
- Labels and Annotation
- Save a Map Document

#

Map Scale and Zoom Tips

Map scale can be defined as the ratio of a distance on the map to the corresponding distance on the ground. A map's scale is often described as a ratio such as 1/24,000 or 1:24,000. This ratio states that 1 unit on the map equals 24,000 units on the ground. In other words, one inch on the map represents 24,000 inches in real world distance. In addition to expressing a map scale as a ratio, the map scale could be described descriptively. For example, with a 1:24,000 map, the scale could be described as "1 Inch = 2,000 Feet". The ratio could also be described with the use of a scale bar. The scale bar allows a user to compare a map measurement and convert to real world coordinates.

In ArcGIS, you can change the scale at which your data are displayed as long as the map units are set correctly. You can enter a scale in the map "Scale Box" (Figure 63) and ArcMap will automatically update the map display.

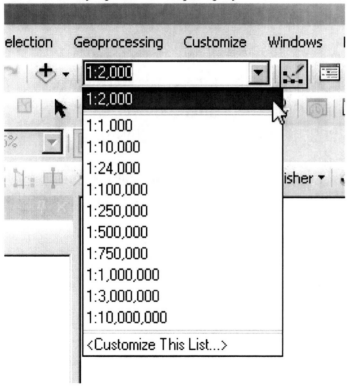

Figure 63: Map Scale Dropdown

1. Launch ArcMap and open the map document, DSD.mxd, found in the "...\ArcBasics\DSD" folder
2. Currently the Map Scale dropdown is grayed out and cannot be changed (Figure 63). You can only change the "Map Scale" when the "Map Units" for the document have not been set.
3. Right-click on the "Layers" data frame and click on "Properties". There are multiple tabs available in the "Data Frame Properties" window (Figure 64).
   - Click on the "General" tab.
   - Change the "Map Unit" dropdown from Unknown Units to Feet.
   - Click "Ok" to close the Data Frame Properties window

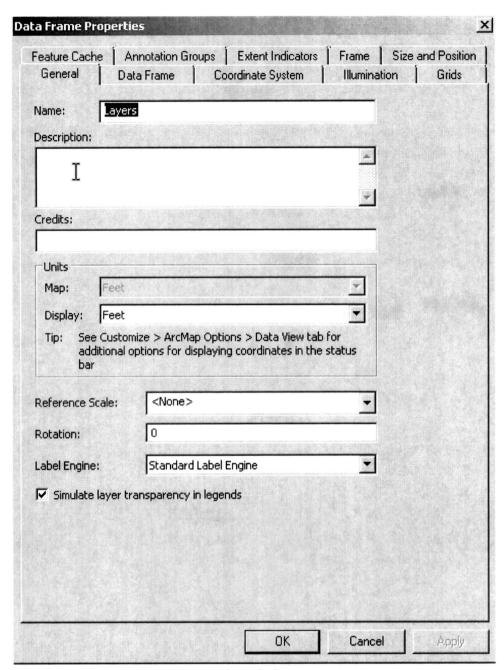

Figure 64: Data Frame Properties window

4. Because the "Map Units" are now set, the user is now able to change the map scale. Change the map scale to 1:5,000. The scale bar should be as follows:

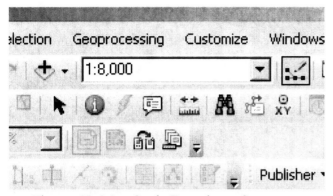

Figure 65: Map Scale Dropdown Set

5. Test various scales. Get a feel for how each new scale affects the map display.

** The scale of a map is often described as small or large scale. Maps that are considered small scale show large areas such as world and large regional maps. For example a map of the United States could be described as small scale. Large-scale maps are those that depict smaller areas in greater detail. A good example would be a detailed map of a small town. **

Besides changing the map scale in the "Map Scale" Dropdown, a user can change the map scale by using the "Zoom In" and "Zoom Out" tools.

Familiarize yourself with the Zoom tools in Figure 66 to change scale in the DSD.mxd.

Figure 66: Zoom Tools

1. Test the different changes in scale when using the "Fixed Zoom In" and "Fixed Zoom Out" buttons.

2. Try the" Full Extent" button found on the Standard toolbar.
3. Right-click on the name of a data layer in the TOC and choosing the "Zoom to Layer" icon.

4. Use the "Go Back to Previous Extent" and "Go to Next Extent" buttons found on the toolbar.

5. Experiment with the "Zoom In" , "Zoom Out ", and "Pan" tools . Use the zoom tools to drag a box around an area of interest.

Become comfortable with the zoom tools before moving to the next section. These are some of the most commonly used functions in ArcGIS. Being able to do basic navigation is the key to learning to master ArcMap.

#

Turning Data Layers On and Off

1. Zoom to the extent of the Trees layer
2. Make sure that the Streets and APSU Buildings layers are displayed on your map. To display a layer, turn it on by clicking on the check box to the left of the layer name in the TOC. Note: A checked layer is On (visible).

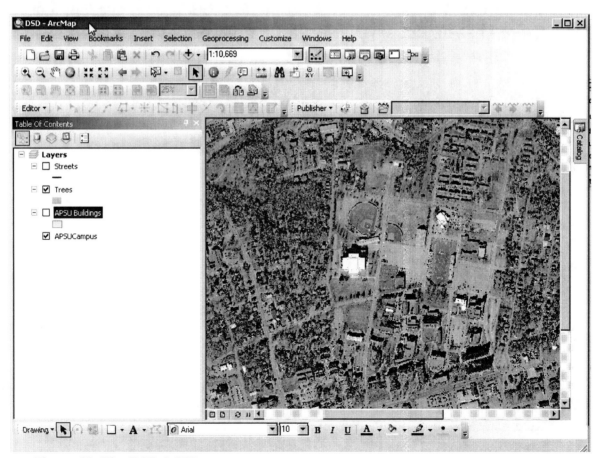

Figure 67: The DSD MXD

#

Ordering Data Layers in the TOC

As you learned in the Intro Section, the ordering of layers in the TOC, dictates the drawing order in ArcMap. Those at the top of the TOC are drawn on top. When the List By Drawing Order tab (located at the bottom of the TOC) is selected, layers are shown in the order in which they will be drawn. One can move a layer up or down in the list but the user cannot change the drawing order of layers when the "List By Source" or "List By Selection" tabs are selected.

1.   Make sure the List by Drawing Order tab is selected at the top of the TOC.

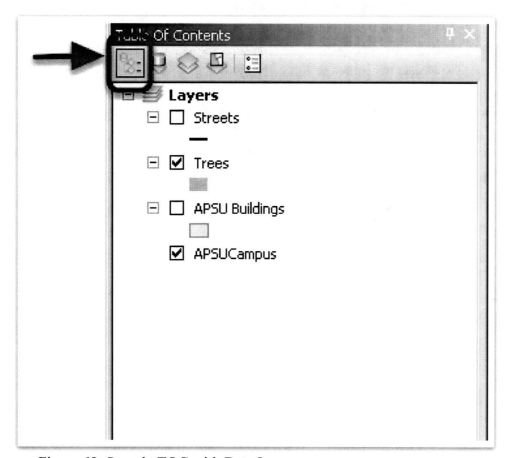

Figure 68: Sample TOC with Data Layers

2.   Click and Hold on the APSU Buildings layer name and drag it to the top of the TOC.
3.   Click and Hold on the APSU Buildings layer name and drag it to the bottom of the TOC.

Notice that when you put the APSU Buildings layer at the top of the TOC, all the other points, lines, and polygons are covered by the Building polygons. When the APSU Buildings layers are moved to the bottom of the TOC (beneath the photo), the polygons are no longer visible. It is good practice to place point layers at the top of the TOC, then line layers. Finally, place polygon layers and photos at the bottom of the TOC. This is the default order when adding layers using the Add Data button in ArcMap.

#

Layer Properties

The Layer Properties dialog allows you to view and/or control all aspects of a layer such as:

- Changing the display name of a layer
- How to draw or symbolize the layer
- How to display selected features
- What data source the layer is based on
- Whether and how to label the layer's features
- Attribute field properties
- What data are joined or linked to the layer

For example, to display and manipulate the properties of the APSU Buildings layer:

1. Right-click the APSU Buildings layer name. In the TOC and select "Properties" from the context "Layer Properties" that appears. You can also double-click a layer name in the TOC to display its "Layer Properties" dialog.

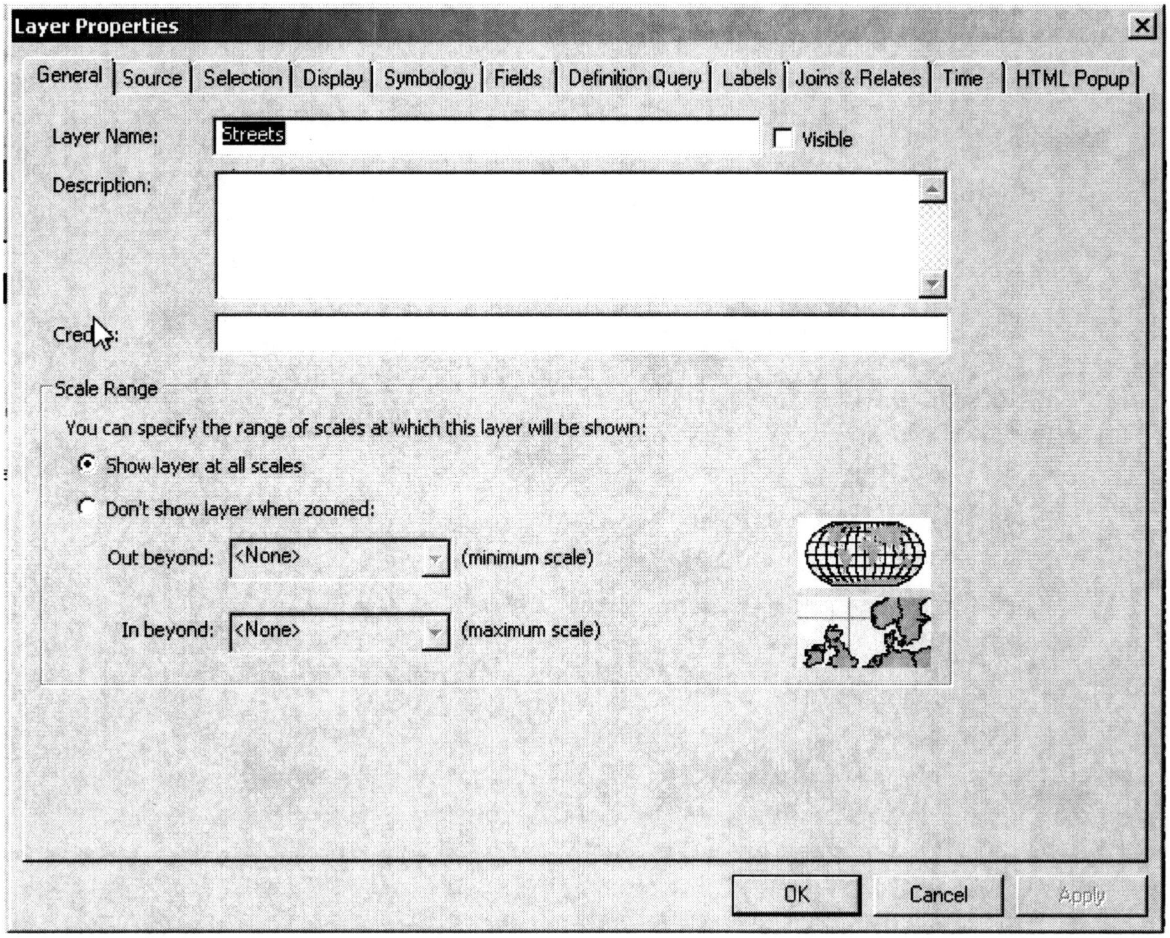

Figure 69: Data Layers Properties

2. Click the General tab.
    - You can change the name of the layer
    - Turn the layer on or off
    - Set a map scale above or below which this layer will not be visible.
3. Click the Source tab (Figure 69) to view information about the source data.

Notice the "Set Data Source..." button that allows you to redirect a layer to another data source. This is particularly useful when a data source has been moved from its original location.

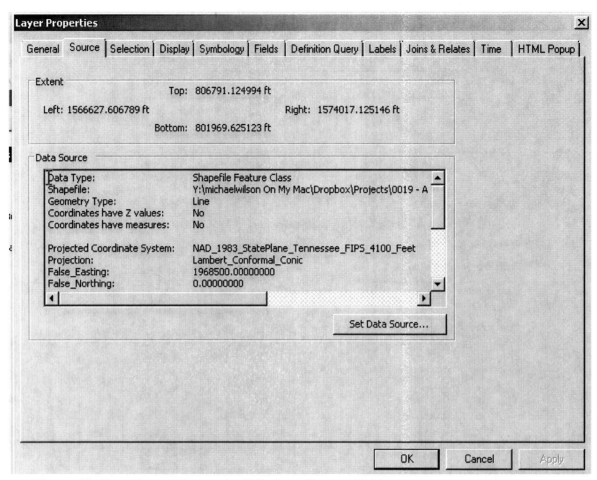

Figure 69: Data Layers Properties Window (Source Tab)

4. The Selection tab allows you to specify a symbol and/or color for displaying selected features.

5. The Display tab allows you to make several changes to the way a layer is displayed:
   - ☐ "Scale Symbology" dynamically (Checked by default)
   - ☐ "Set layer Transparency". This is particularly helpful when displaying polygons over photos. The polygons can be made semi-transparent, allowing you to see the underlying photos.
   - ☐ Set field to use for "MapTips".
   - ☐ Create a "Hyperlink" field. Linked fields open other documents, images, or websites when clicked with the hyperlink tool.
   - ☐ "Feature Exclusion", are features not to draw in the display or layout.

6. The "Symbology" tab is used to control classification and display of the layer's features.

7. Click the "Fields" tab to see the name, type, length, precision, and scale of each attribute (field) in the layer's attribute table. Additionally, you can check and uncheck fields to make them visible or invisible.

It is important to note that when you uncheck a field and then export the data, the unchecked field WILL NOT be in the exported dataset.

8. The "Labels" tab allows you to control various aspects for displaying labels.
9. Click each of the other tabs to view or manipulate other layer properties. When you are finished, close the "Layer Properties" dialog.

#

Set Display Units and Measure Distance on the Map Display

In ArcMap, you can measure the distance between two points using the Measure Tool located on the Standard Toolbar. Display Units is the data frame property that controls what measurement unit is used to report distance measurements by default. The following steps illustrate how to set display units for a data frame and then how to use the measure tool to measure the distance between two points on the map display.

1. First, set the "Display Units" for the Layers data frame. Right-click on the "Layers" data frame name at the top of the TOC.
2. Choose "Properties" at the bottom of the context menu.
3. Under the "General "Tab, note that "Map Units "are set to Feet and "Display Units" are set to Unknown Units.
   As was explained in the Intro Section, map units must be set to the measurement units in which the data are stored or projected (e.g. UTM-17 (meters), NAD83 (meters), State Plane (Feet)).
4. Change the "Display Units" by selecting Feet from the dropdown list (Figure 70). Click "OK".

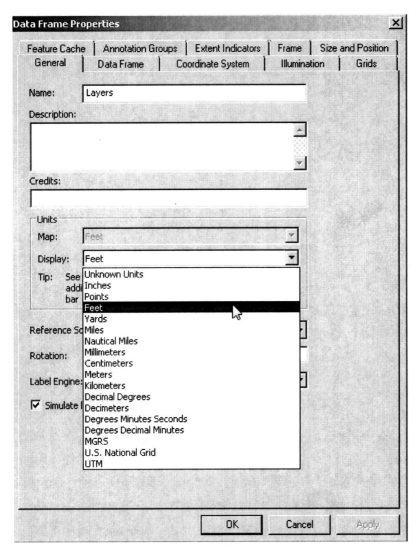

Figure 70: Setting Display Units

5. Select the "Measure" tool from the toolbar to measure the distance between two Buildings.

6. Click once on an Emergency Phone then double-click on another Buildings.

   **Note that the Measure Tool will snap between features. **

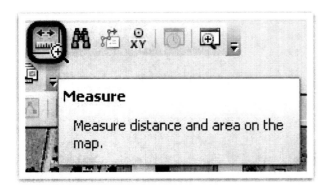

Figure 71: Taking a Measurement

7. Look at the Measure Tool Dialog to see the distance, in feet, between the two Buildings.
8. Try setting the Display Units to meters and then measure the distance between two other Buildings.

Figure 72: The Measure Tool Dialog

As noted above, the "Measure Tool" dialog provides the user with the ability to perform a variety of measurements. This dialog will allow you to:

* Measure a Line
* Measure a Feature
* Measure a Polygon
* Keep a sum of measures
* Measure Areas (and set units)
* Measure Distances (and set units)

#

Display an Attribute Table

1. Make sure the Parcels layer is turned on and that it is placed at the bottom of the TOC. Right-click the APSU Buildings layer in the TOC.
2. From the context menu that opens, choose the "Open Attribute Table" option (Figure 73).

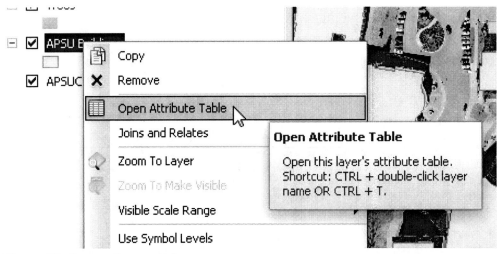

Figure 73: Layer Context Menu

3. Use the scroll bars on the right and bottom of the table window to scroll through the fields and records in the table. Notice the caption at the bottom of the table, Records (0 out of 85 Selected) – this table contains a total of 85 records.

FID	Shape	BUILDING_N	STORIES	HEIGHT	BUILDINGID	EYB	BLDGSQFT	Shape ▲
44	Polygon	Hand Village #200	2	19	A01032	2003	14575	46
45	Polygon	Hand Village #300	2	19	A01033	2003	14575	32
46	Polygon	Hand Village #400	2	19	A01034	2003	14575	33
47	Polygon	Hand Village #500	2	19	A01035	2003	14575	34
48	Polygon	219 Home Ave	1	9.5	A8007	0	0	20
49	Polygon	Meacham Apts #1	2	19	A00701	0	0	5
50	Polygon	Meacham Apts #2	2	19	A00703	0	0	66
51	Polygon	Meacham Apts #3	2	19	A00704	0	0	60
52	Polygon	204 Castle Hgts	2	19	A8003	0	0	19
53	Polygon	206 Castle Heights	2	19	A8004	0	0	20
54	Polygon	208 Castle Heights	2	19	A8005	0	0	18
55	Polygon	212 Castle Heights	2	19	A8006	0	0	25
56	Polygon	214 Castle Heights	2	19	A0094	0	0	22
57	Polygon	Hand Village #700	2	19	A01037	2003	14575	32
58	Polygon	Hand Village #800	2	19	A01038	2003	14575	33
59	Polygon	Ellington	3	28.5	A0008	1951	41966	70
60	Polygon	Cross Hall	2	19	A0035	1966	34818	69
61	Polygon	Killbrew Hall	2	19	A0055	1969	37572	63
62	Polygon	Castle Hgts 219	2	19	A0080	1940	2360	18
63	Polygon	Castle Hgts 217	2	19	A0091	0	0	19

85 (0 out of 85 Selected)

APSU Buildings

Figure 74: Parcels Attribute Table

Sort Attribute Table Records

1. In the Attributes of Parcels table, right-click on the heading for the field (column) named [BUILDING_N]. The following menu containing options for manipulating data in the table appears:

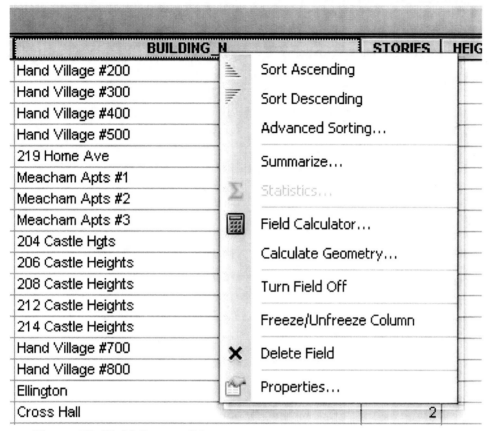

Figure 75: Field Context Menu

2. From this menu click the "Sort Ascending" button. Notice the records in the table are now sorted in alphabetical order according to the values in the [BUILDING_N] field, allowing you to find records with a specific class value more easily.
3. Try the "Sort Descending" button with the [STORIES] field and notice how the records are sorted.

Select Attribute Table Records

1. Use the "Select Elements" tool to click the gray box to the left of the first field. Because both the polygons and the records are the same data, ArcMap highlights both the selected table record and the feature associated with that record in the map display, as shown in Figure 76.
2. Hold down the <Ctrl> key and select some additional records. To select consecutive records, click on the first record and drag the cursor down the field until you reach the

last record you wish to select. Notice the updated information regarding number of records selected is displayed in the Attribute table status bar.

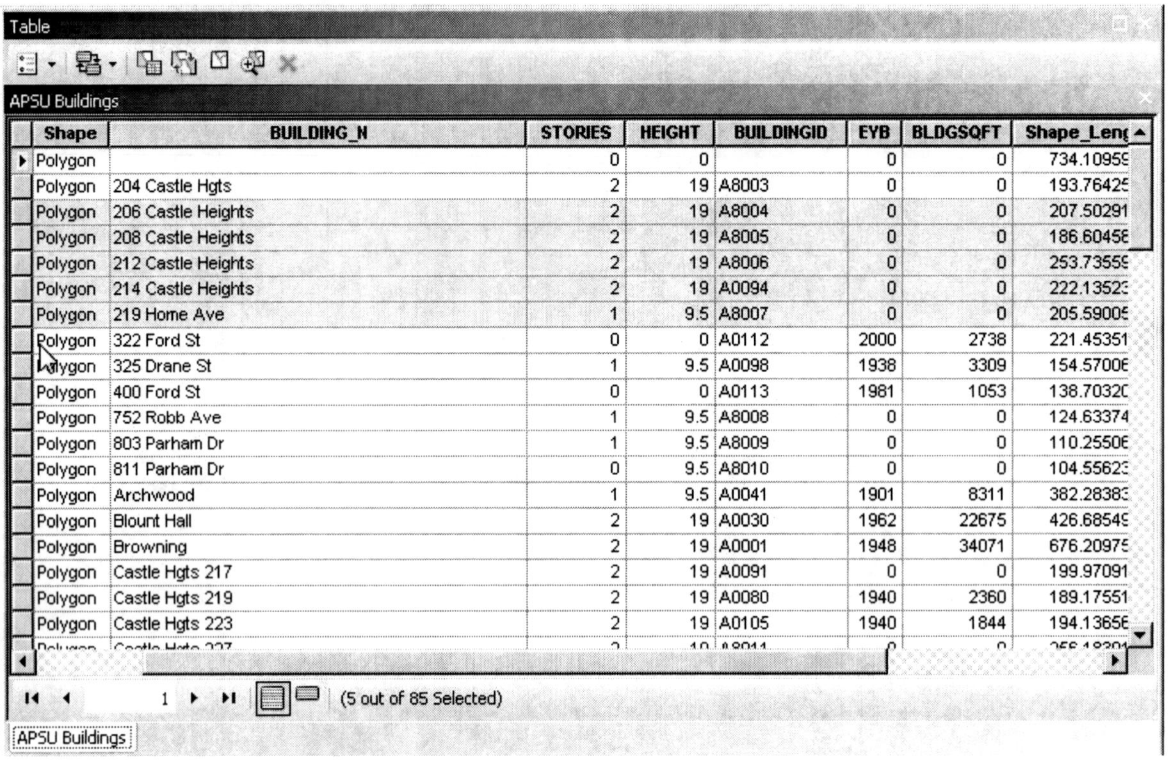

Shape	BUILDING_N	STORIES	HEIGHT	BUILDINGID	EYB	BLDGSQFT	Shape_Leng
Polygon		0	0		0	0	734.10959
Polygon	204 Castle Hgts	2	19	A8003	0	0	193.76425
Polygon	206 Castle Heights	2	19	A8004	0	0	207.50291
Polygon	208 Castle Heights	2	19	A8005	0	0	186.60458
Polygon	212 Castle Heights	2	19	A8006	0	0	253.73559
Polygon	214 Castle Heights	2	19	A0094	0	0	222.13523
Polygon	219 Home Ave	1	9.5	A8007	0	0	205.59009
Polygon	322 Ford St	0	0	A0112	2000	2738	221.45351
Polygon	325 Drane St	1	9.5	A0098	1938	3309	154.57006
Polygon	400 Ford St	0	0	A0113	1981	1053	138.70320
Polygon	752 Robb Ave	1	9.5	A8008	0	0	124.63374
Polygon	803 Parham Dr	1	9.5	A8009	0	0	110.25506
Polygon	811 Parham Dr	0	9.5	A8010	0	0	104.55623
Polygon	Archwood	1	9.5	A0041	1901	8311	382.28383
Polygon	Blount Hall	2	19	A0030	1962	22675	426.68549
Polygon	Browning	2	19	A0001	1948	34071	676.20975
Polygon	Castle Hgts 217	2	19	A0091	0	0	199.97091
Polygon	Castle Hgts 219	2	19	A0080	1940	2360	189.17551
Polygon	Castle Hgts 223	2	19	A0105	1940	1844	194.13658
Polygon	Castle Hgts 227	2	19	A8011	0	0	255.18291

(5 out of 85 Selected)

APSU Buildings

Figure 76: Selected Records

3. To clear or unselect the records, click on the "Options" button in the bottom right of the table window and then select "Clear Selection". Click on "Switch Selection" to inverse the selected records.

4. If you wish to unselect selected records in all layers, click on the "Selection" menu item in the menu bar and choose "Clear Selected Features" (as shown below). This will unselect all selected features, and consequently all selected records, in all layers.

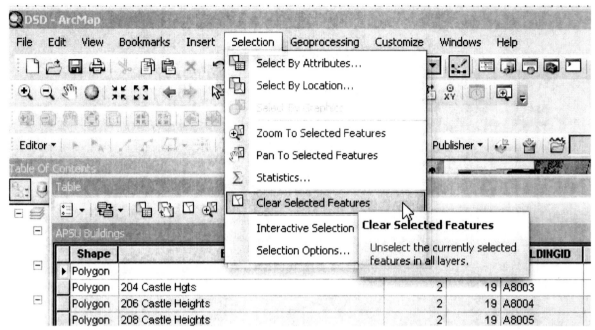

Figure 77: Selection Menu

5. Close the Attributes of APSU Buildings table.

#

Select Features Interactively

In ArcMap, there are several methods to select features. As mentioned in the previous section, you can select features by selecting their associated attributes. Features can also be selected interactively through several different methods. Before starting to select features, you should set the selection parameters. Setting the appropriate parameters will insure that you select the features that you think you are selecting and to increase the efficiency of the selection process. These parameters include a set of selection options, selectable layers, and selection method.

Set Selection Options

Click on "Selection" in the menu to open the "Selection Options" dialog, click on "Options" (see Figure 77).

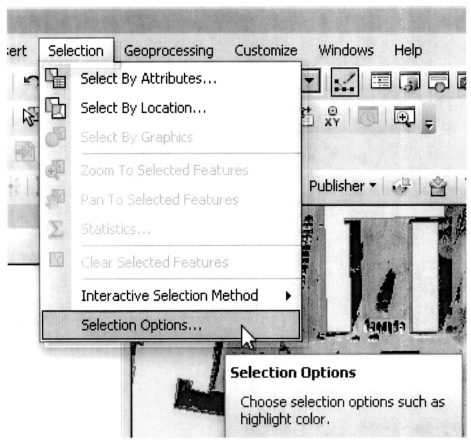

Figure 77: Selection Menu

The "Selections Options" dialog (Figure 78) contain numerous settings:

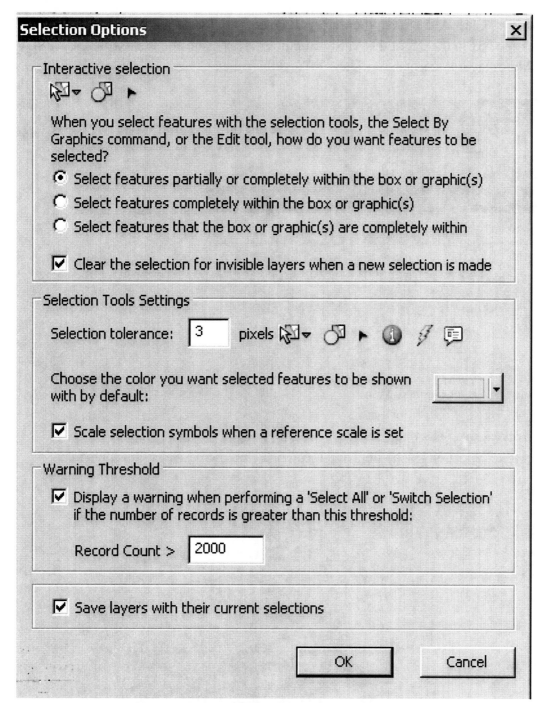

Figure 78: Selection Options Dialog

Review the headings and individual parameters that you can set using the "Selection Options" dialog (Figure 78). For now, you do not need to change any of the default settings. Click "Cancel" to close the "Selection Options" dialog.

Set Selectable Layers

1. Click the "List by Selection" tab on the TOC.

2.   Click the "Clear All" button to unselect all layers.
3.   Click the box to the left of the APSU Buildings layer name and make sure the APSU Buildings layer is selectable (Figure 79) and then click Close.

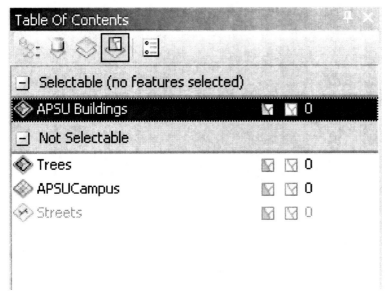

Figure 79: Layer Context Menu

Set Interactive Selection Method

1.   Click the "Selection" menu item and then click "Interactive Selection Method":

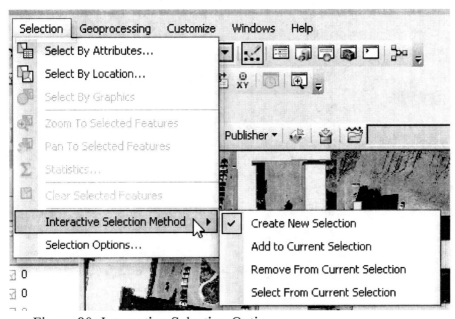

Figure 80: Interactive Selection Options

2. There are several selection methods available (Figure 80), for this demonstration use the default selection, "Create New Selection".

Select Individual Features from the Map Display

1. Use the "Zoom In" tool to set the map scale where you can easily see individual Building boundaries.
2. Click on the "Select Features" tool in the Tools toolbar (Figure 81). When you move the cursor to the map display, it will look like this:

Figure 81: Selection Tool and Types of Selection

3. Click on an APSU Building polygon. To select more than one APSU Building polygon, hold down the <Shift> key and click on additional polygons.
4. Select several APSU Building polygons on the map display.
5. Next, open the attribute table for the APSU Building layer: right-click on the APSU Building layer name in the TOC and select "Open Attribute Table". The dialog at the bottom of the window shows the number of Records selected (Figure 82).

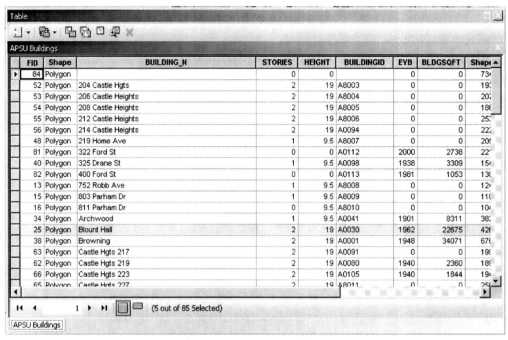

FID	Shape	BUILDING_N	STORIES	HEIGHT	BUILDINGID	EYB	BLDGSQFT	Shape
84	Polygon		0	0		0	0	73
52	Polygon	204 Castle Hgts	2	19	A8003	0	0	19
53	Polygon	206 Castle Heights	2	19	A8004	0	0	20
54	Polygon	208 Castle Heights	2	19	A8005	0	0	18
55	Polygon	212 Castle Heights	2	19	A8006	0	0	25
56	Polygon	214 Castle Heights	2	19	A0094	0	0	22
48	Polygon	219 Home Ave	1	9.5	A8007	0	0	20
81	Polygon	322 Ford St	0	0	A0112	2000	2738	22
40	Polygon	325 Drane St	1	9.5	A0098	1938	3309	15
82	Polygon	400 Ford St	0	0	A0113	1981	1053	13
13	Polygon	752 Robb Ave	1	9.5	A8008	0	0	12
15	Polygon	803 Parham Dr	1	9.5	A8009	0	0	11
16	Polygon	811 Parham Dr	0	9.5	A8010	0	0	10
34	Polygon	Archwood	1	9.5	A0041	1901	8311	38
25	Polygon	Blount Hall	2	19	A0030	1962	22675	42
38	Polygon	Browning	2	19	A0001	1948	34071	67
63	Polygon	Castle Hgts 217	2	19	A0091	0	0	19
62	Polygon	Castle Hgts 219	2	19	A0080	1940	2360	18
66	Polygon	Castle Hgts 223	2	19	A0105	1940	1844	19
65	Polygon	Castle Hgts 227	2	19	A8011	0	0	25

(5 out of 85 Selected)

APSU Buildings

Figure 82: Selected APSU Building in Attributes

6. Press the "Selected" button next to the Record count (Figure 83).

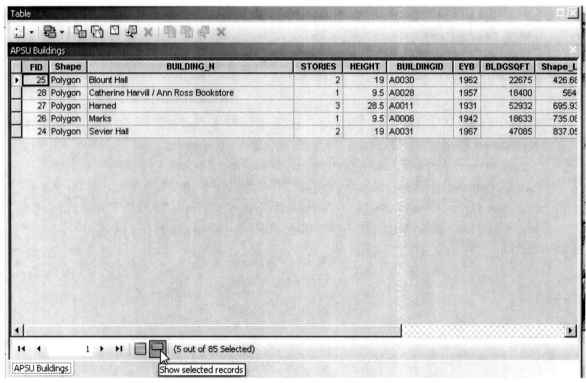

FID	Shape	BUILDING_N	STORIES	HEIGHT	BUILDINGID	EYB	BLDGSQFT	Shape_L
25	Polygon	Blount Hall	2	19	A0030	1962	22675	426.68
28	Polygon	Catherine Harvill / Ann Ross Bookstore	1	9.5	A0028	1957	18400	564
27	Polygon	Harned	3	28.5	A0011	1931	52932	695.93
26	Polygon	Marks	1	9.5	A0006	1942	18633	735.08
24	Polygon	Sevier Hall	2	19	A0031	1967	47085	837.05

(5 out of 85 Selected)

APSU Buildings   Show selected records

Figure 83: Only Selected Parcels in Attributes

7. The window sorts out all other records, displaying only the selected attributes. Press the "All" button at the bottom of this window to redisplay all the records.
8. Use the "Selection" menu item to unselect all records ("Clear Selected Features").
9. Close the Parcels attribute table.

Select Groups of Features from the Map Display

To select groups of features:
- ☐   Select features using the lasso tool
- ☐   Select via an onscreen graphic.

1. Turn off all layers except the Trees and APSU Buildings layers. Right-click on the APSU Buildings layer and select "Zoom To Layer" to display the full extent of this layer and center it on the screen.

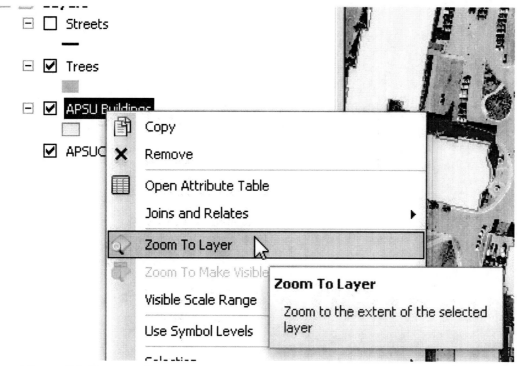

Figure 84: Zoom to Layer

2. Make Trees the only selectable layer. The data display should now look like Figure 85.

Figure 85: Set Trees as Selectable

3. Click on the Select Features tool. Click and drag a rectangle around the Trees in the southwestern portion of the APSU Campus. When you release the cursor button, the Trees inside the rectangle that you formed with the cursor are selected (Figure 86).

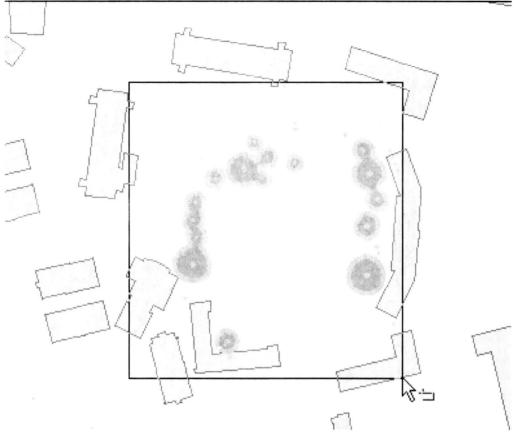

Figure 86: Trees Selected

4. Clear the selected features.
5. To use the "Select By Graphics" method, you must first draw a graphic around the features you wish to select. If it is not already enabled, turn the Draw toolbar on. Click the dropdown arrow to the right of the "Draw", a rectangle button and select the "New Polygon" icon, Figure 87:

Figure 87: Select New Polygon in Draw Toolbar

6. Move the cursor to the map display and draw a polygon that encloses several Emergency Phones. When you close the polygon, it will probably display with a fill that obscures the Phones. When you remove the polygon graphic (in step 8) after performing the selection, the features will be selected. Click on the "Selection" menu item and then on "Select By Graphics". In the map display, you will see the selected features inside the graphic you drew (Figure 88).

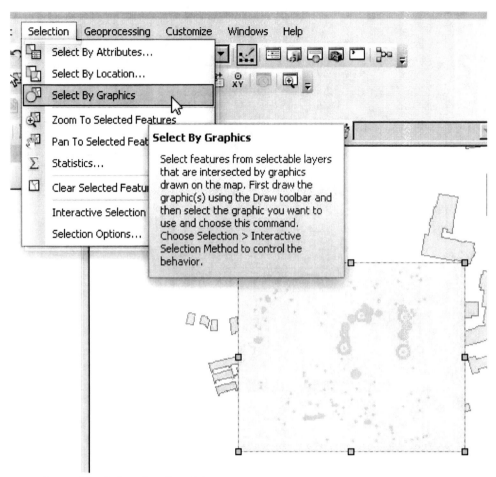

Figure 88: Select by Graphic

You have the ability to change the fill style for the polygon graphic. Right-click inside the polygon graphic and select "Properties", then select the "Symbol" tab and" Fill Color". Choose "No Color" and the selected point features become visible.

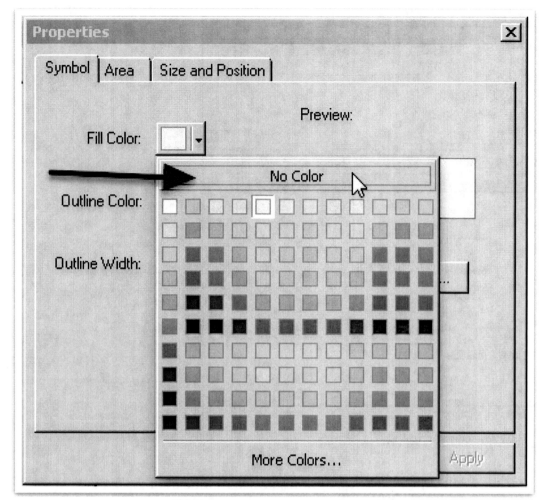

Figure 89: Change Polygon Fill

1. Open the attribute table for the Trees layer and click the Selected button below the table to see the records that are selected.
2. Close the Attributes of Trees table, clear selected features, and delete the polygon graphic. To delete the graphic, select it with the Select Elements tool and press the <Delete> key.

#

Use the Identify Tool to See the Attributes of a Feature

1. Turn on the Streets layer.
2. Click on the Identify tool located on the Tools toolbar. Once you click on the Identify tool the appearance of the cursor changes.
3. Using the Identify tool, click on a Street to see its attribute information. The Identify window displays the attribute data associated with the feature.

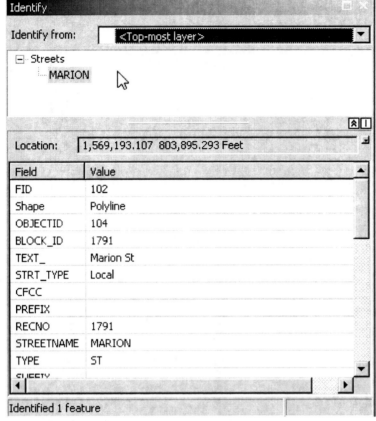

Figure 90: Identify Window

By default the Identify Results window displays the attributes of the feature in the <Top-most layer>. If you want to view the attributes of a feature in a different layer, use the dropdown list of Layers located at the top of the window to choose the appropriate layer.

1. Notice that the attribute window also shows the coordinates for the point you clicked on. Change the value of the coordinates to Decimal Degrees (Figure 91). Although the value is changed in the Identify window, the coordinate system for the layer remains the same.

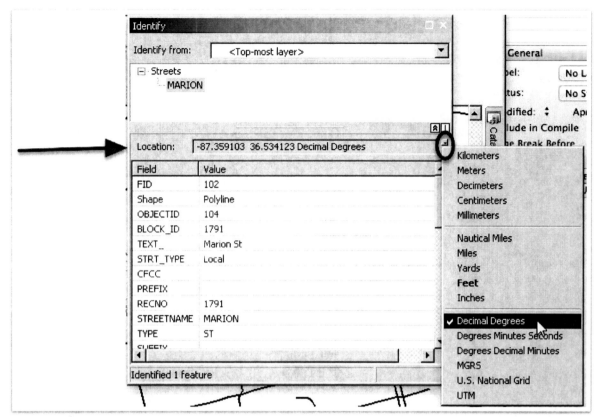

Figure 91: Identify Window Location

1. Close the Identify Results window.

#

Labels and Annotation

Labels and annotation are the two main ways that a user can display text on a map. A label's text and position are generated dynamically according to a set of placement rules. You have a limited set of options to change the placement roles. You are unable to select a label and change the label position on the map. When you use annotations, the position, text string, and display properties are stored with each annotation independently. When deciding between using labels or annotations, if placement matters, use annotation. ArcGIS supports the display and conversion of various types of annotation.

In ArcMap, labels for features can be based on any attribute values. In the next example, you will label each Street based upon Street Name. The field to be used in the attribute table is "LABEL".

1. Verify the Streets layer is on.
2. Right-click on the Streets layer, in the TOC and select properties from the context window. In the Layer Properties window the Labels tab (Figure 92).

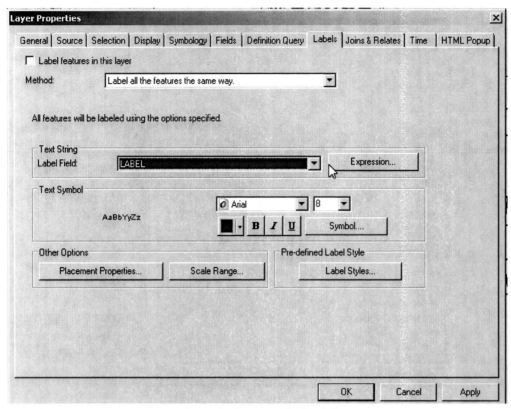

Figure 92: Layer Properties Window - Labels Tab

3. Check the box next to "Label features in this layer". In the method drop-down select "List Label All Features The Same Way".
4. Select the field, "LABEL", in the label field drop-down list (Figure 93).

Figure 93: Select the field to use as a label

5. Before closing the "Layer Properties", review the scale range and label styles options.
6. Click "OK" on the "Layer Properties" window. After the window closes, notice the Streets are now labeled by Street Name.
7. To quickly turn the labels off, right-click on the Streets layer, and uncheck the "Label Features" option in the context window (Figure 94).

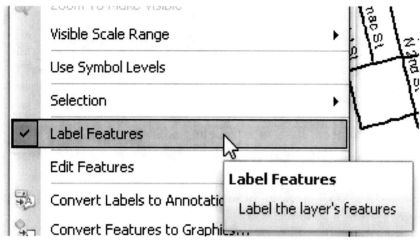

Figure 94: Turn off Labels

Zoom in and zoom out on the map, notice that the labels stay the same size regardless of scale. In order to have the labels change size as the map changes scale, use the following steps:

1.   In the Standard toolbar, click in the map scale box and enter "24,000" and press enter. The map is now set to 1:24,000 scale.
2.   Right-click on the layers data frame in the TOC, from the context menu select reference scale > set reference scale. The reference scale for the map is now 1:24,000.
3.   Use the zoom in tool to zoom to a small area containing several Streets. Notice how the labels changed scale as the map changes scale (Figure 95).

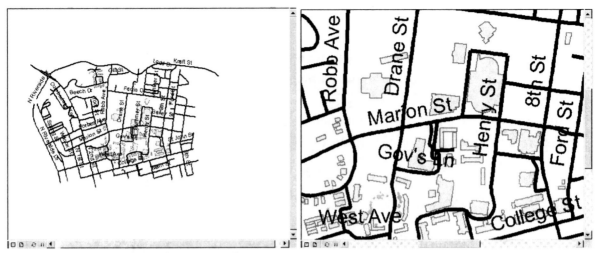

Figure 95: Set Reference Scale

ArcMap provides a number of advanced labeling options:

1. Label classes – allow the user to define classes of features and specify labeling properties for each class.
2. Label expressions – allow you to control how strings are formatted from feature attributes.
3. Text symbol – controls how text appears on screen.
4. Placement properties – allows you to specify the placement of the labels in respect to the position of the feature.
5. Label properties – allow you to specify which features are labeled and the order the labels appear on the map. This is particularly important when numerous features and layers are being labeled.

Labels are not editable, meaning that you cannot select, move, or change the display of individual labels. In contrast, annotation is editable text.

Convert Labels to Annotation

As stated earlier, if you need to control how a label is placed and how they are displayed, you should use annotations. In the following example, you will learn how to convert labels to annotation.

1. Verify the Streets layer is turned on and the labels we created previously are displayed.
2. Right-click on the Streets layer in the TOC and select convert labels to annotations from the context menu (Figure 96).

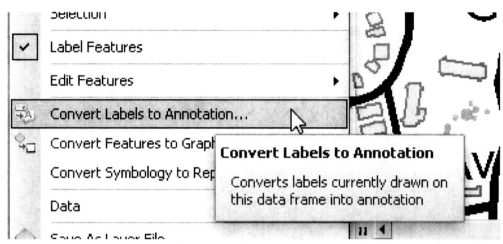

Figure 96: Convert Labels to Annotation

3. Once you click on the convert labels to annotations, the convert labels to annotations dialog will open. In the dialog, you are presented with two options for storing the annotations, "In the map" and "In a database" (Figure 97 - 1). "In the map" stores the annotations as part of the map document. "In a database" stores the annotations in a standard geodatabase annotation feature class. Click "In the map".

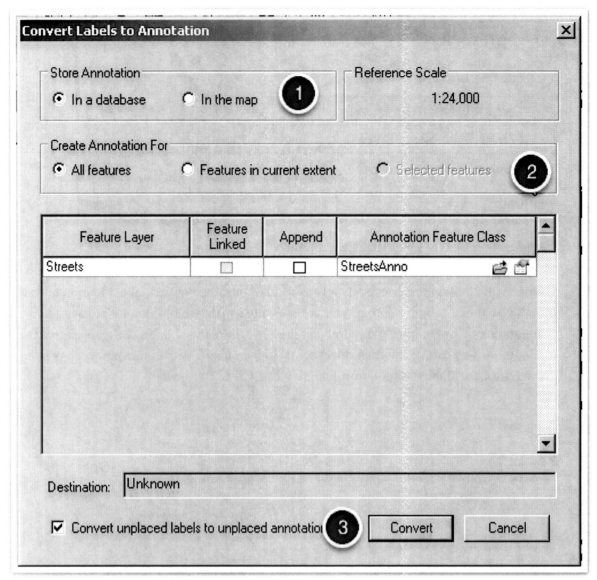

Figure 97: Convert Labels to Annotation Dialog

4. The "Convert labels to annotations" dialog has three options for which features to "Create annotations for" (Figure 97 - 2). These options are "All features", "Features in current extent", and "Selected features". Click on "All features".

5. Review your selections and click "Convert" (Figure 97 - 3).

For more information regarding labels and annotations, please see "ArcGIS desktop help".

#

Map Tips

Map tips are a way for you to display selected information on the map as you move the cursor over features. You define an attribute field that will "Pop up" when you pause the mouse pointer over a feature in the ArcMap data display window. This is a quick way to see the name of a feature or some other piece of information about it without having to use the Identify tool.

1. In the TOC, right-click the APSU Buildings layer name and then click "Properties" from the context menu that appears.
2. Click the "Display" tab (Figure 98).
3. Click the "Display Expression" dropdown arrow and select the BUILDING_N attribute field.

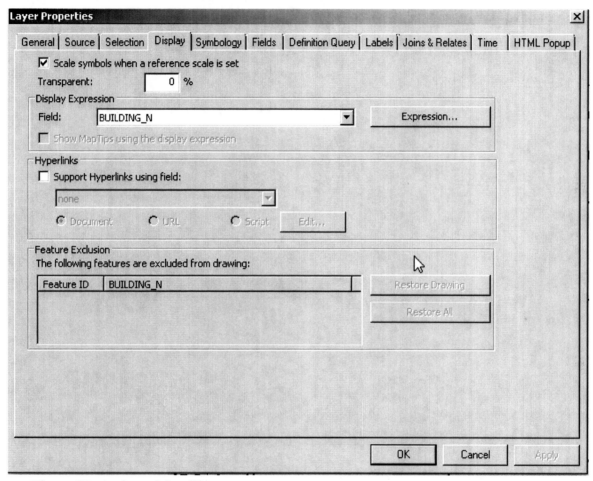

Figure 98: Activate Map Tips

4. Click on the "Show MapTips" using the display expression and then click "OK".
5. Move the mouse pointer over a plot in the data view map display to see the map tip.

\#

Save a Map Document

To save the map document, click on the "File" menu item, and then click "Save as". ArcMap will open a traditional Windows "Save" dialog. Save your map as DSD_user.mxd in the "..\ArcBasics\DSD\Data" folder.

#

DSD Exercise

Use the DSD MXD to answer the following questions:

1. Who owns the largest property on the Parcels data layer?
2. In the APSU Buildings data layer, sort the BUILDING_N field in the attribute table and select the building with then name of "Sundquist Science". On which intersection is the building located?
3. What is the distance, in feet, from this building to the "Morgan University Center"?

Chapter 5 - Selecting Features

Introduction

ArcMap provides powerful tools for users to select features by attribute or by location. To select by attribute, a user must specify a set of criteria. For example, select all parcels that are greater than 2 acres. It is possible to build both simple and complex queries. When selecting by location, you are comparing the spatial relationships of layers. For example, select all emergency phones within 500 feet of in an APSU building. In addition to selecting features, attribute and location queries can be used to limit features that are shown on a map.

#

Real World Application

Based on the GIS definition, GIS is an information system, a database, of spatial information. As with any database, attributes in GIS can be queried. A user of GIS is expected to query and utilize the GIS as any database user. Knowing how to query and select data based upon attribute values is critical. Querying by attribute is only part of the power of GIS. A true GIS system can also query based upon location. A spatial query allows the user to select objects based upon their location as a specific point, a relationship, or in reference to another layer.

#

Objectives

In this section, you will learn how to do the following:

1. Using the find tool to find features that match user-defined criteria.
2. Selecting features based on attribute values.
3. Selecting features based on location.
4. Exporting selected features, to create new feature data sets.
5. Using definition queries to display a subset of features.

#

Find Features

The "Find" button in the standard toolbar allows you to locate features based upon attribute information. The "Find" tool is similar to the find function in Microsoft Word.

1. Open the Select.mxd located in the "...\ArcBasics\Select" folder.

2. Click the [binoculars icon] ("Find" tool) in the "Tools" toolbar. This will open the "Find" dialog (Figure 99).

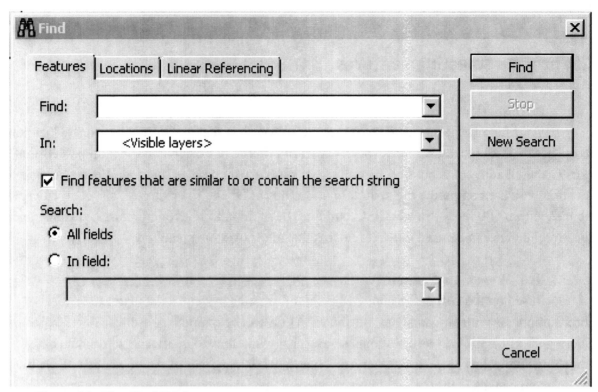

Figure 99: The Find Dialog

3. Type the text "Be" in the text box labeled "Find:"
4. In the "In:" drop-down select "Soils".
5. Verify that the "All fields" radio button is selected.
6. Click the "Find" button in the dialog to show all values in the "Soils" layer that contain the value "Be".
7. Notice in the results window, that any field that contained the search term were returned. Notice, also, that the search is not case-sensitive.
8. If you right-click on one of the results, a context menu will appear. Click on "Zoom to" (Figure 100). You can also click on the "Flash" item to show the feature. Notice that there are a variety of functions in the context menu; take a minute to explore the functionality.

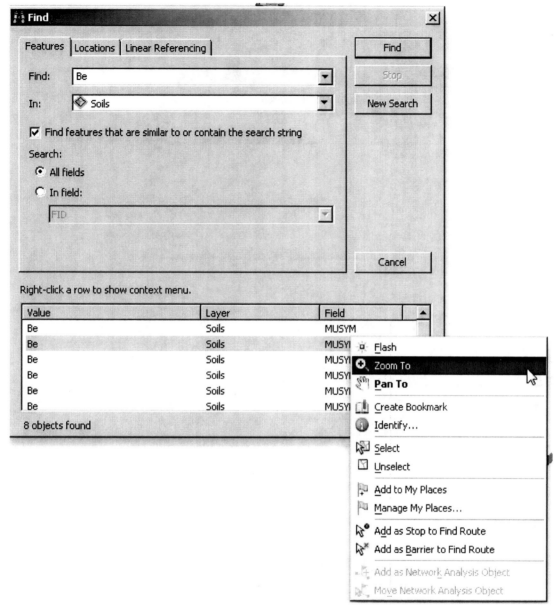

Figure 100: Find Soils

#

Select by Attributes: Simple Queries

In the following example, you will learn how to use ArcMaps "Select by attributes" function. "Select by attribute" allow you to select features based on values in the features attribute table.

1. Make sure the "Select.mxd" is open.
2. Open the attribute table for the Parcels data layer. To open the attribute table, Right-click on the Parcels layer name in the TOC and click "Open Attribute Table".
3. In the ArcMap menu bar click on "Selection" menu item and then click on "Select by Attributes". This opens the "Select by Attribute" dialog (Figure 101).

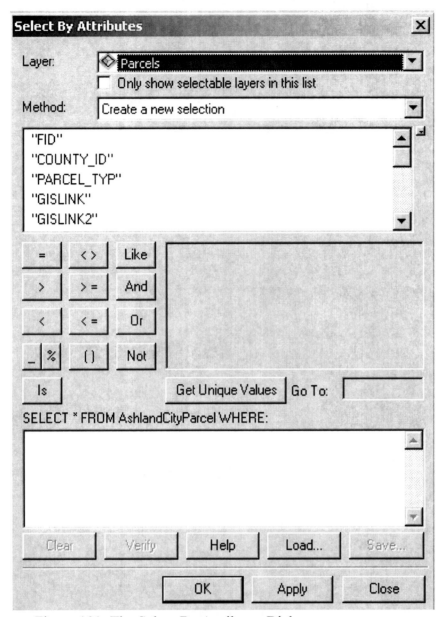

Figure 101: The Select By Attributes Dialog

4. From the "Layer" drop-down, click on the parcels layer.
5. From the "Method" drop-down list, click on "Create a new selection".
6. In the "Fields" list box double-click on "CALC_ACRE". Make sure the field name is added to the expression box.
7. In the expression box, type the > operator and type 50.
8. Click the "Verify" button to validate the query in make sure you have used the proper syntax. Once you have verified the expression click "OK".
9. Click the "Apply" button and ArcMap will execute the query. Close the "Select by Attribute" dialog and review the results on the map (Figure 102).

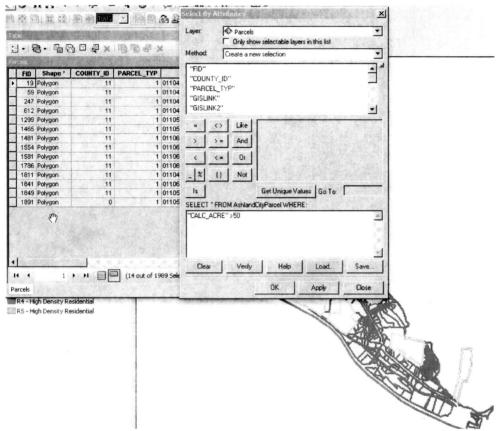

Figure 102: Select Parcels greater than 50 Acres

As you can see, you have selected all the parcels greater than 50 acres. With ArcMap, it is simple to add additional features to the selection. Use the following steps, to add features to the current selection.

1.   Reopen the "Select by attribute" dialog. In the expression box change 50 to 25.
2.   In the "Method" drop-down list, click on "Add to Current Selection", click the apply button. As you can see, ArcMap has expanded the number of selected features to include parcels greater than 25 acres (Figure 103).

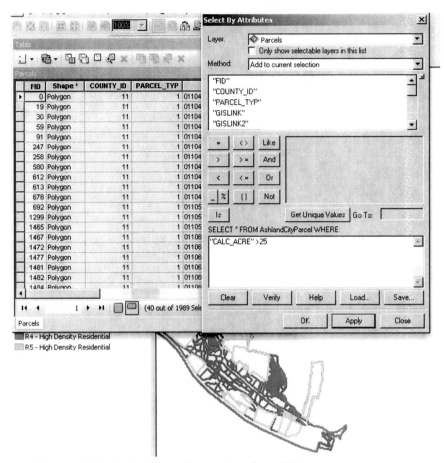

Figure 103: Select Parcels greater than 25 Acres

3. Click the "Clear" button, to clear the values in the expression box. The "Clear" button does not clear the selected attribute records.

Using the selection tools and ArcMap, you can make a selection based on previous selected records. For example, one can select a subset of the parcels.

1. Reopen the "Select by attribute" dialog from the "Method" drop-down. Click on "Select from current selection".
2. In the "Fields" list, double-click on the "OWNER" field. Verify that the "OWNER" field is added to the expression box.
3. From the operators, click on the "like" button.
4. In the expression window, type 'B%'. When searching for a text value, be sure to use the " single quotes.
5. Click the "Verify" button to make sure the query syntax is correct.
6. When you click the "Apply" button, ArcMap will select all owner names that begin with the letter B and have 25 or more acres (Figure 104).

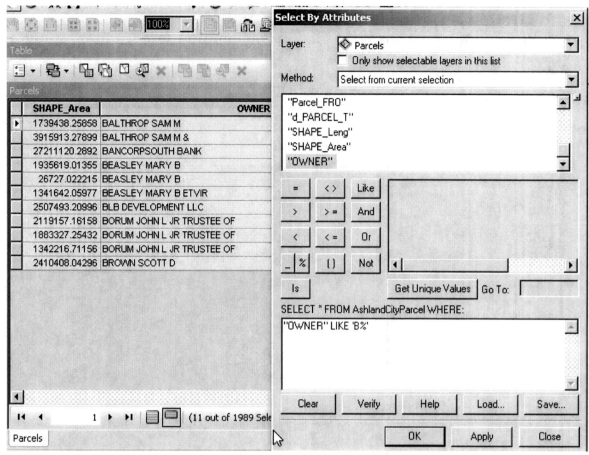

Figure 104: Select Parcels greater than 25 Acres with an Owner name starting with 'B'

7. In the attribute window, click on the "Options" button and then click on "Clear selection". No records should now be selected. Selected records can also be cleared by clicking on "Selection" menu item and then by clicking on "Clear selected features".
8. Close the "Select by Attribute" dialog.

Notice: you can also launch the "Select by attribute" dialog, from the "Attributes of parcels" table by clicking on the "Options" button and then clicking on "Select by attribute" (Figure 105).

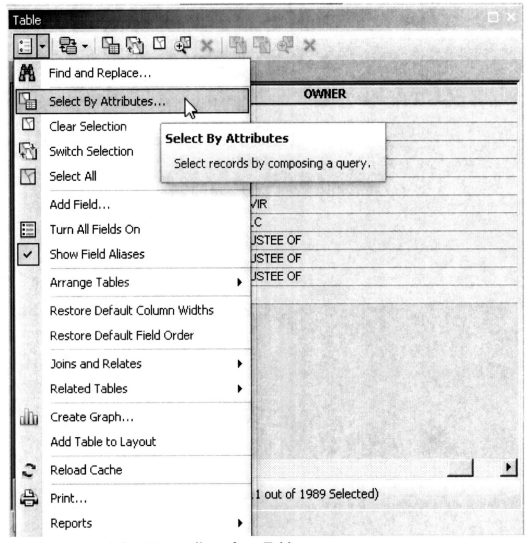

Figure 105: Select By Attribute from Table

#

Select by Attributes: Complex Queries

As we have seen previously, you can build simple queries to select features. In many cases, you will need to build complex queries to make selections. The following example will show you how to build a complex query to select features. We will select all parcels that are on map '55C' and have an Owner Name starting with 'W'.

1. Open the "Parcels" attribute table by Right-clicking on the layer name in the TOC and selecting "Select by attribute".
2. In the "Select by attribute" dialog in the "Method" drop-down list, click on "Create a new selection".
3. In the "Fields" list box double-click on "Parcel_MAP" and verify the field name is added to the expression box.
4. Click on the = operator button.

5.   Type '55C' in the expression box. The expression should look something like "Parcel_MAP" = '055C' (Figure 106).

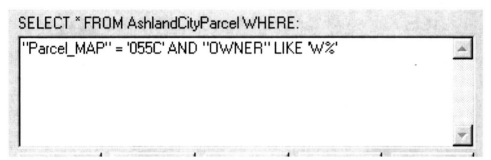

SELECT * FROM AshlandCityParcel WHERE:

"Parcel_MAP" = '055C'

Figure 106: Select "Parcel_MAP" = '055C'

6.   Click on the "And" operator.
7.   From the "Fields" list box, double-click on "OWNER".
8.   Click on the LIKE operator.
9.   In the expression box type 'W%'. The complete expression, should look like "Parcel_MAP" = '055C' AND "OWNER" LIKE 'W%' (Figure 107).

SELECT * FROM AshlandCityParcel WHERE:

"Parcel_MAP" = '055C' AND "OWNER" LIKE 'W%'

Figure 107: Select "Parcel_MAP" = '055C' AND "OWNER" LIKE 'W%'

10. Click on the "Verify" button, once you have verified the query, click "Apply". Notice the selected records in the attribute table (Figure 108).

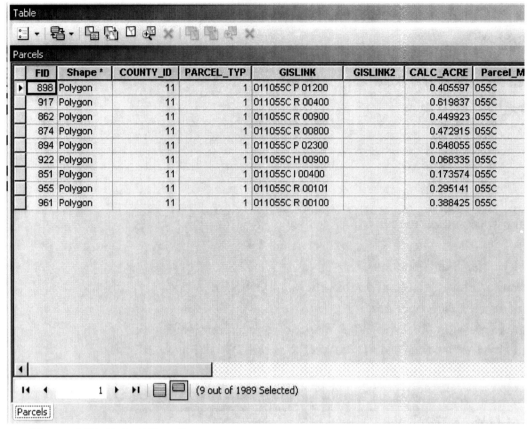

Figure 108: Select Results

#

Save and Load a Query Statement

One of the underused features in ArcMap is the ability to save structured query language, known as SQL. SQL statements can be saved and loaded into other map documents for reuse.

1.   Open the "Select by attribute" dialog and create a simple query statement using the skills you learned previously.
2.   In the lower right of the "Select by Attribute" dialog, click the "Save" button to save the query statement as an expression file (*.exp).
3.   Save the expression in the "...\ArcBasics\Select\Files" folder.
4.   Test loading and saving various query expressions (Figure 109). Use the "Clear" button to clear the expression box.

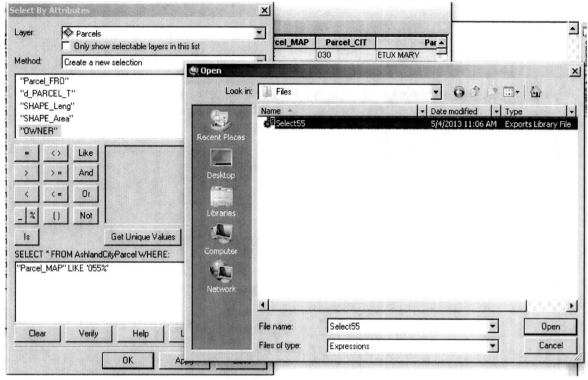

Figure 109: Load SQL Expression

5. Close the "Select by Attribute" dialog.

#

Show and Clear Selected Records in an Attribute Table

1. In the "Attributes of Parcels" table you should have several records selected. To show with the selected records go to the bottom of the attribute table and click on the "Selected" button (Figure 110). Only the selected records are now visible.

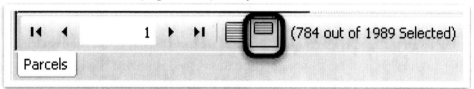

Figure 110: Show Selected Records

2. At the bottom of the hatch window, click on the "All" button to display all records (Figure 111).

Figure 111: Show All Records

3. To unselect records, click on the "Options" button in the attribute window, and then select "Clear Selection" (Figure 112).

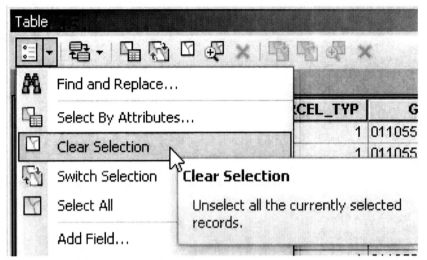

Figure 112: Clear Selected Records Table Options

You can also use the "Clear Selected" from the Tools toolbar (Figure 113).

Figure 113: Clear Selected Records Tools toolbar

4. Close the "Attributes of parcels" table.

\#

Select by Location

In the following exercise, you'll learn how to select features based upon their spatial relationship with other future layers. The "Select by Location" tool is one of the most powerful functions in ArcMap. This function requires that you have at least two layers loaded.

1. Make sure "Select.mxd" is open. In the TOC, make sure that the "City Owned Parcels" layer is turned on.
2. From the "Selection" menu, click on "Select By Location".
3. In the "Select by Location" dialog, click on the "Selection Method" drop-down list and select "Select features from".
4. In the layers list, check the box next to the "Parcels" layer.
5. From the "Source layer" drop-down list, select "City Owned Parcels".
6. Check the "Apply Search Distance" checkbox.

7. Enter a search distance of 100 feet.

8. Click Apply. ArcMap will select all "Parcels" within 100 feet of any "City Owned Parcels".

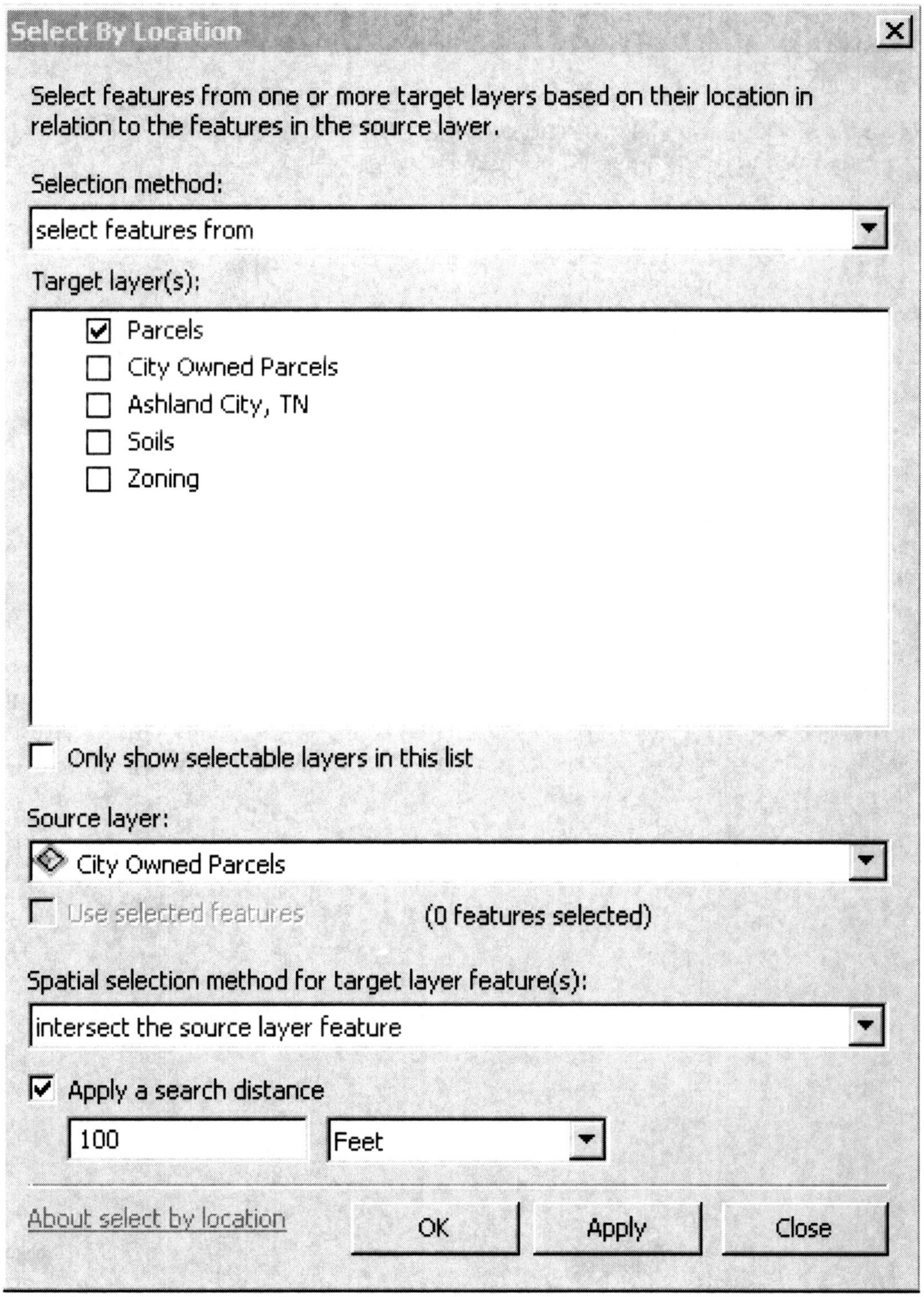

Figure 114: Select By Location

9. Close the Select.mxd without Saving.

In the next set of steps, we will use the "Select by Location" function to find all APSU Emergency Phones that are located within the boundary of a Parking lot and are within 100 feet of an APSU building.

1. Open the Select2.mxd located in the "...\ArcBasics\Select" folder.
2. In the menu, click on "Selection". Then click on "Select by Location".
3. In the "Select by Location" dialog, make sure the "Select Method" drop-down shows "Select Features From".
4. In the "Target Layers" list box, be sure to check "Emergency Phones".
5. In the "Source layer" drop-down box, click on "APSU Buildings".
6. Check the box next to "Apply Search Distance" and enter 100 feet.
7. Click "Apply". Notice that ArcMap has selected all emergency phones that were within 100 feet of a building.
8. In the "Select by Location" dialog, make sure the "Select Method" drop-down shows "Select From The Currently Selected Features".
9. In the " Target Layers" list box, be sure to check "Emergency Phones".
10. In the "Source Layer" drop-down box, click on "Parking".
11. Uncheck the box next to "Apply Search Distance".
12. In the "Spatial Selection Method" drop-down box, click on "Are Completely Within The Source Data Layer".
13. Click "Apply". Notice that ArcMap has selected all emergency phones that were within 100 feet of a building and in a parking lot (Figure 115).
14. Click "Close" to close the "Select by Location".

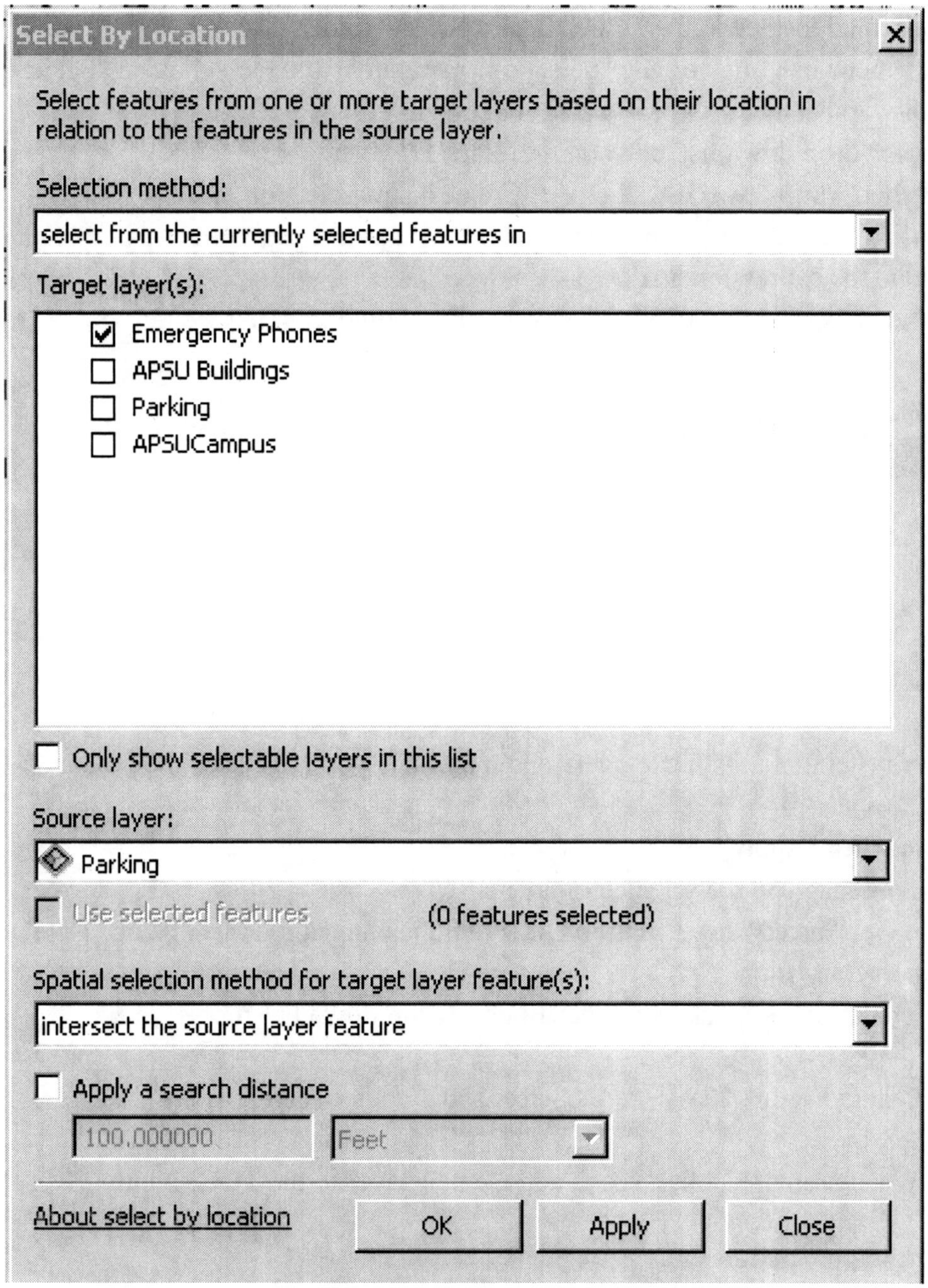

Figure 115: Select By Location - Emergency Phones

#

Export Selected Features

When you are selecting features in ArcMap, you can export the results to a new shape file or geo-database feature class. The ability to export features allows you to work within a subset of data while leaving the original data intact. Creating a subset is also important when selecting records from large data sets. For example, you may have a layer of all Parcels in a City, but for your project you are interested in just the parcels on one map. You can easily select just the parcels on one map and create a new layer.

1.   Open the Select.mxd located in the "...\ArcBasics\Select" folder.
2.   In the ArcMap menu bar, click on the "Selection" menu item and then click on "Select by attributes". The "Select by attributes" dialog will open.
3.   From the "Layer" drop-down list, click on the "Parcels" layer.
4.   From the "Method" drop-down list, click on "Create a new selection".
5.   In the "Fields" box, double-click on "Parcel_MAP". Make sure the "Parcel_MAP" field name is added to the expression box.
6.   In the expression box, type " = '055'". The expression should look like:

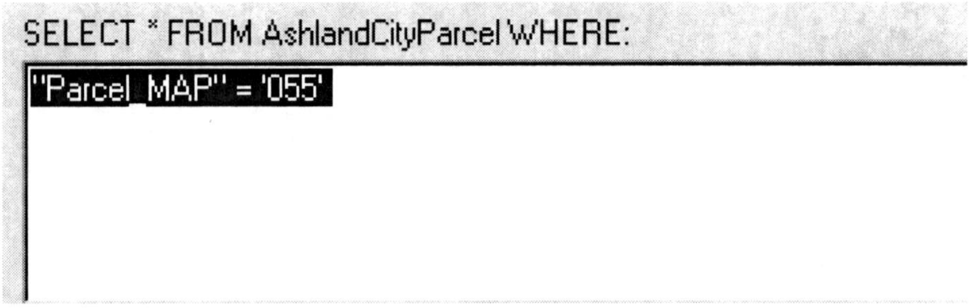

```
SELECT * FROM AshlandCityParcel WHERE:
"Parcel_MAP" = '055'
```

Figure 116: Select Map 055 Parcels

7.   Click the "Verify" button to validate query and make sure you have used the proper syntax.
8.   Click "OK" and then "Apply".
9.   All the Parcels on map 055 are selected in the map.
10. Right-click on the "Parcels" layer in the TOC. From the context menu, click on "Data" and then "Export Data".
11. In the "Export data" dialog, make sure the "Export" drop-down list shows "From Selection".
12. Change the "Output Feature Class:" to "..\Select\Data" folder, name the layer Parcels055.shp.
13. Click "OK". When prompted, click yes to add the new layer to the TOC (Figure 117).

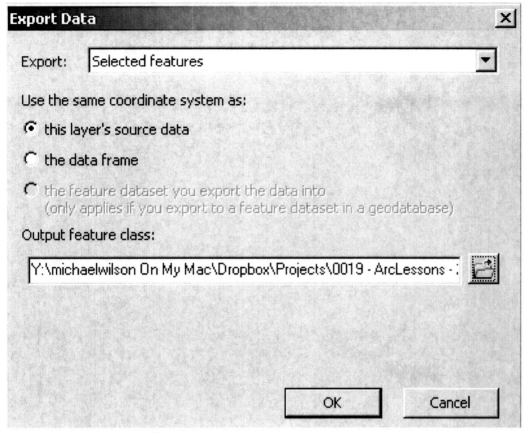

Figure 117: Export Data

14. There should now be two layers in the TOC, "Parcels" and "Parcels055". Take a moment to explore each data set.

                                        #

Display a Subset of Features in a Layer: Using a Definition Query

In ArcMap, a definition query is used to create a subset of features in a layer. The difference between a definition query and data layer that was created based on saved selection is that the definition query is still the same data as the original data set. It is only displaying the selected records. Definition queries are ideal in situations where a master data set is being edited often and a definition query allows you to work with a single data set.

1. In the TOC, turn on the "Zoning" layer.
2. Right-click on the parcels layer name and click on "Properties" from the context menu.
3. Click the "Definition query" tab.
4. Click on the "Query builder" button.
5. In the "Field" list box, double-click on the "Zoning" field. The field name should appear in the expression box.
6. From the list of operators, click on the LIKE button.
7. Type the value 'R%'.

8.   The expression box should look like "Zoning" LIKE 'R%'.

9.   Click on the "Verify" button to validate the query syntax.

10.   Click on the "OK" button (Figure 118).

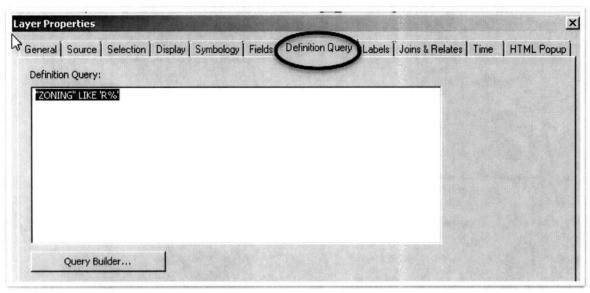

Figure 118: Definition Query Tab

11. Notice that the map now displays only residential zoning. However, all original data remains intact.

#

Similar to a definition query, you can easily create a layer based upon a selection. A selection layer is a subset of a dataset. The power of the subset is that as you edit the master dataset, the changes are automatically in the selection set. In the following exercise, you will create a selection layer.

1.   Open the Select.mxd located in the "...\ArcBasics\Select" folder.

Figure 119: Selection Tool

2. In the "Tools" toolbar, use the "Selection" tool to select some Parcels (Figure 119).
3. Once the Parcels are selected, right-click on the name of the Parcels layer in the TOC.
4. From the context Menu, click on "Selection", and the "Create Layer from Selected Features" (Figure 120).

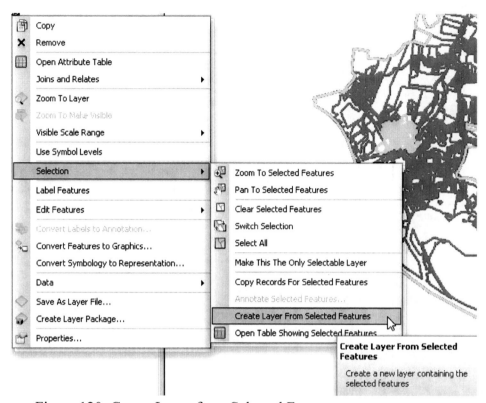

Figure 120: Create Layer from Selected Features

5. The new layer, "Parcels Selection" will appear in the TOC. Open the "Layer Properties" dialog, and click on the "General" tab.
6. In the "General" tab, change the "Layer Name".
7. Click "Apply" and "OK".

#

Exercise 1

You are working for a county and an industry is looking to build a facility in your county. The industry is interested in properties that are at least 200 acres and within 500 feet of a gas line.

• Open a new map document.
• Add the data layers from the ...\ArcBasics\Select\Data folder to the map:
  ☐ Parcels.shp
  ☐ GasLines.shp

- Create a map that shows all parcels within 500 feet of a gas line and at least 200 acres. Hint: Use a definition query to show only parcels that are at least 200 acres.

1. How many 200-acre parcels are within 500 ft. of a gas line?
2. How many of these parcels have a "LANDVAL" over $1,000,000? How many are less than $6,000,000?

Chapter 6 - Symbolizing Data

Introduction

When using ArcMap, you have the ability to load data from a variety of sources. These various layers are then represented on a map. Although using the default layer symbology will create a passable map, your job as a professional is to create an informative map that maximizes the data being conveyed. In order to present a professional map, you can use the power of ArcMap to symbolize the data.

#

Real World Application

Take a look at the following maps. Both display the same data. Which conveys the data in the best way?

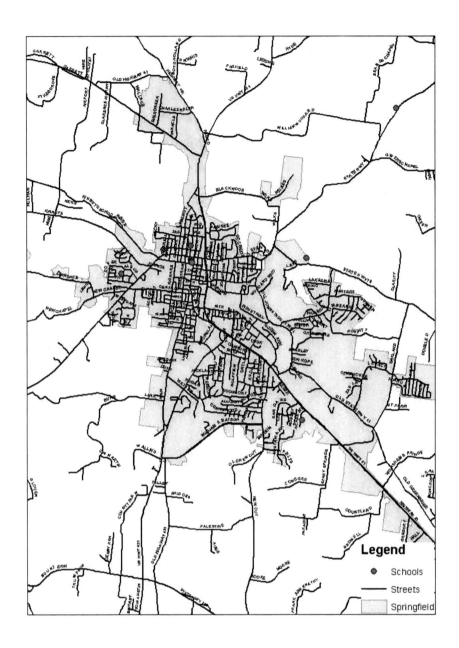

Figure 121: School Maps with Default Symbols

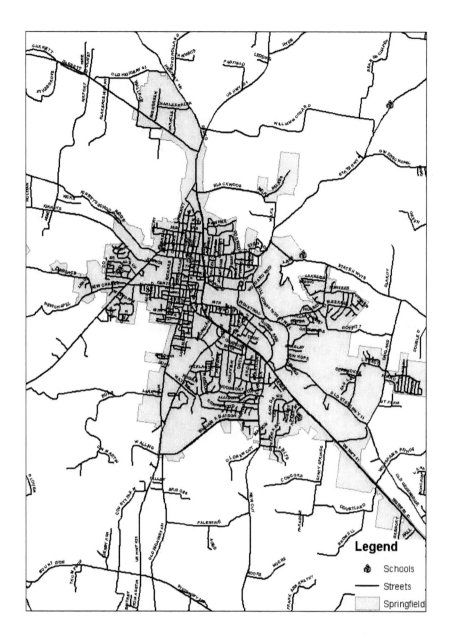

Figure 122: School Maps with User Defined Symbols

Notice that the user-defined symbols on the second map (Figure 122) are more informative. Your job when making maps is to maximize the info on the map.

#

Objectives

In "Symbolizing Data", you will learn how to do the following:

- Symbolize Layers
  - Points
  - Lines
  - Polygons

- Display a Layer Based on Categorical Attribute Data
- Display Layers based on Data Ranges
- Use and create a Layer File

#

Symbolizing a Point Layer

A powerful feature in ArcMap is the ability to display layers in a number of different ways. In the following example, we will use symbolize data in various ways.

As you load data into ArcGIS, you have the ability to symbolize each layer in the map. In this exercise, we will look at how to symbolize point data.

1. Launch ArcMap and open the map document, "Symbol.mxd", found in the "...\ArcBasics\Symbol" folder.
2. Use the Add Data button to add the following layers located in the "...\ArcBasics\Symbol\SHP" folder to the map:

  - Schools
  - Streets
  - Springfield
  - WaterBodies

Notice that each layer is loaded with a default symbol set. In the example below (Figure 123), both the City Limits and the Water Bodies layers loaded with almost the same color.

Figure 123: Layers loaded with Default Symbols

3. There are two methods for changing symbols. Symbols can be changed via the TOC or the Layer Properties Dialog.

   ☐   Via the TOC, Single-Click on the Schools symbol (Figure 124). The Symbol Selector Dialog will open.

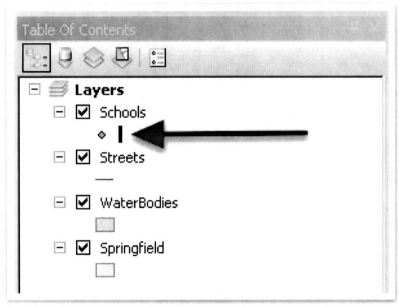

Figure 124: Double-Click on the Symbol in the TOC

☐ Right-Click on the Layer Name in the TOC, and in the Context Menu click on Properties. Go to the "Symbology" tab and click on the Point symbol (Red Arrow in Figure 125) and the Symbol Selector Dialog will open.

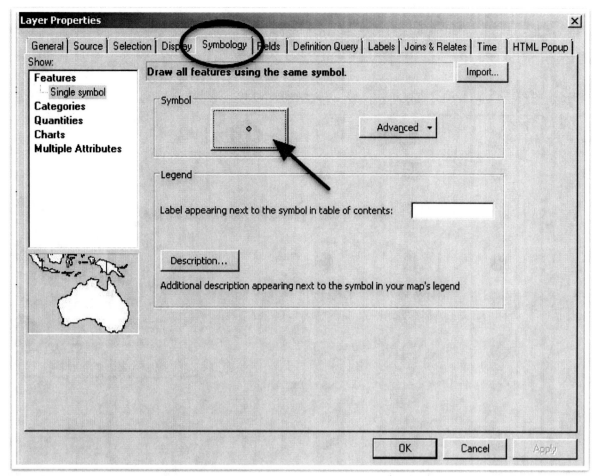

Figure 125: The Symbology Tab

1.   Take a look at the various controls in the Symbol Selector Dialog (Figure 126).

        (1) - Symbol Preview

        (2) - Symbol Search

        (3) - Symbol List

        (4) - Symbol Color Selector

        (5) - Size Selector

        (6) - Advanced Symbol Editor

        (7) - Style References

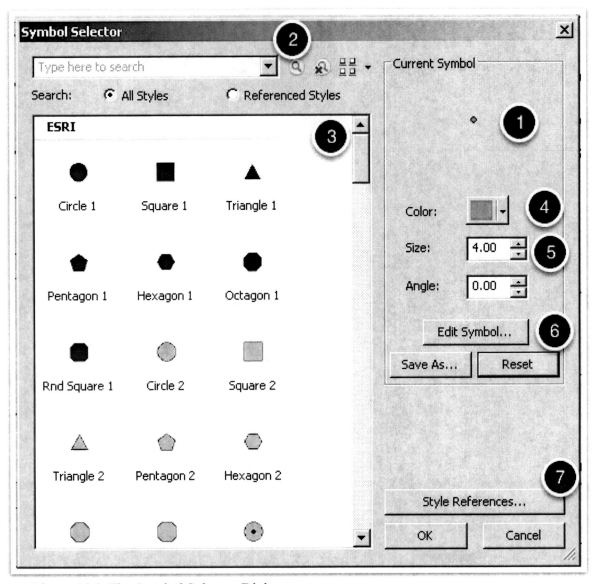

Figure 126: The Symbol Selector Dialog

1.   In the Symbol Search, type "School". Look through the available results (Figure 127) and select a symbol.

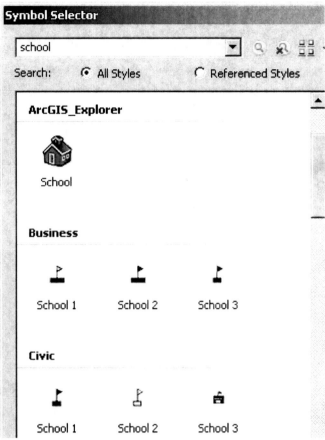

Figure 127: School Symbols

2. Notice that once you select a symbol, it appears in the Symbol Preview (Figure 128).

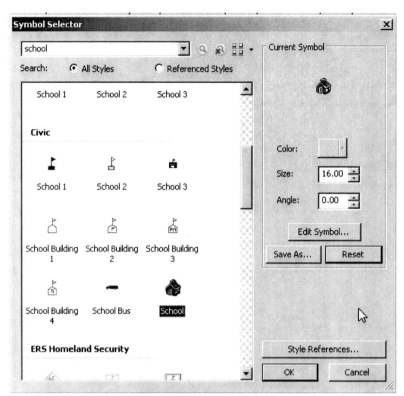

Figure 128: Symbol Selected

3. Edit the symbol size as appropriate.
4. If you close the Symbol Selector and do not like the look of the symbol, follow the steps above and change as needed.
5. Test the various controls in the Symbol Selector.

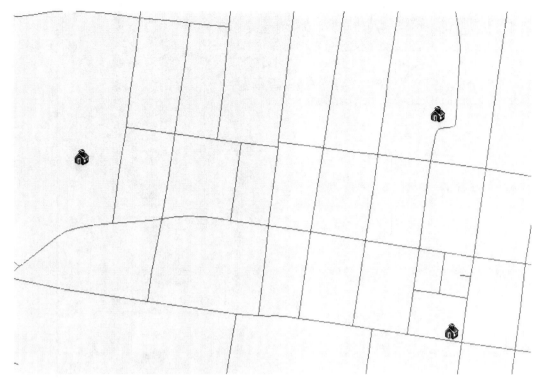

Figure 129: School properly symbolized

6. Once you are satisfied with the look of the school symbol (Figure 129), click "OK" and "Apply", proceed to the next exercise.

#

Symbolizing a Linear Layer

In the following exercise, you will learn to symbolize the linear layer, "Streets".

1. Open the Symbol Selector for the Streets layer via the TOC or the Layer Properties dialog.

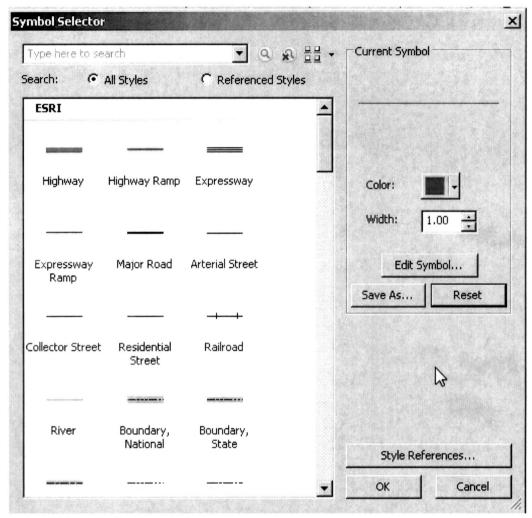

Figure 130: Line Symbols

2. Notice that the available line symbols (Figure 130) are slightly different from the point symbols.
3. The size selector will change the line width.
4. Once you have the have symbolize the streets layer, go to the next exercise.

#

Working with Transparency

There are 2 methods for setting the transparency of a layer, through the Display Tab in the Layer Properties and by the "Effects" Toolbar.

1. In the TOC, right-click the Springfield layer name and then click Properties from the context menu that appears. The Layer Properties Dialog will open.
2. Click the Display tab (Figure 131).

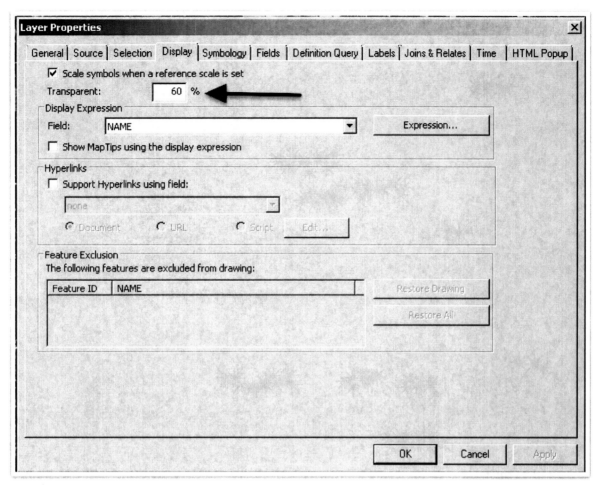

Figure 131: Set Transparency

3. Set the Transparency to 60%.
4. Click "Apply" and "OK".

From the "Effects" Toolbar:

1. Click "Customize" then "Toolbars" and "Effects" on the Main Menu. The "Effects" toolbar appears (Figure 132).

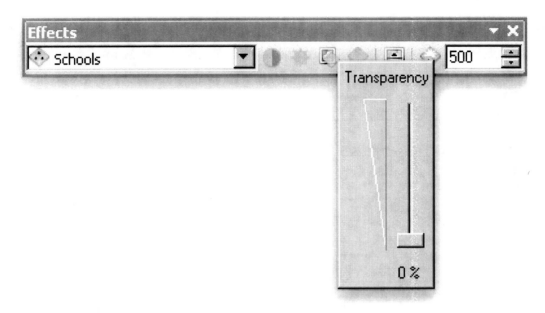

Figure 132:Effects Toolbar

2. Click the layer drop-down arrow and select the layer that you want to appear transparent.
3. Click the "Adjust Transparency" button and drag the slider bar to adjust the transparency.

#

Symbolizing a Polygon Layer

1. Open the Symbol Selector for the WaterBodies layer via the TOC or the Layer Properties Dialog.

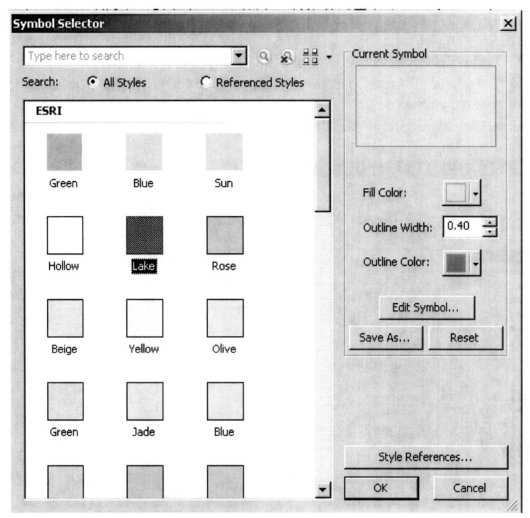

Figure 133: Polygon Symbols

2. The general layout of the Symbol Selector Dialog is similar to what we have seen previously. Notice the Color selector for the Fill and the Outline colors (Figure 133).
3. Take a moment to test various fills and outline types.
4. Change the symbol for the WaterBodies layer to the "Lake" symbol.
5. Save the map and proceed to the next exercise.

#

Display a Layer Based on Categorical Attribute Data

In the following example, you will use unique values in the attribute table to symbolize the parcel layer.

1. Verify the Parcels layer is on. Right-click on the layer name in the TOC and choose properties.
2. In the "Layer Properties" dialog, click on the "Symbology" tab.

3. In the list on the left side of the dialog, select "Categories" and then select "Unique Values".
4. In the "Value field" drop-down, select the "Pt" field (property type).
5. Click on the "Add all values" button. All of the values from the "Pt" field (property type) should now be visible.
6. Click on "Apply" and then click "OK" to apply the changes to the map.

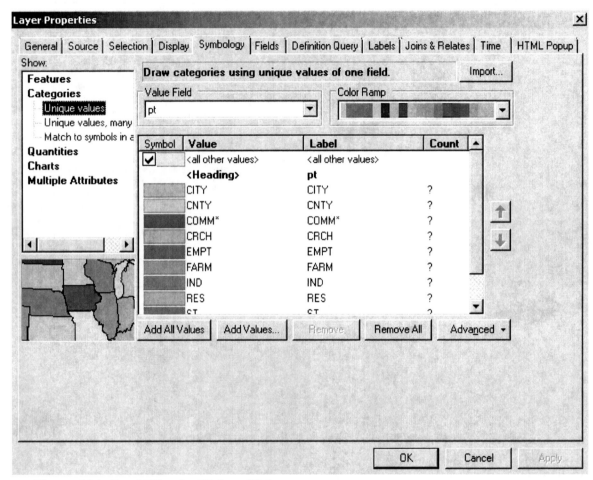

Figure 134: Symbolize by Unique Value

\#

Displaying Layers based on Data Ranges

1. Right-click on the Parcels layer and click on Properties in the Context menu.
2. Click the Symbology tab on the Layer Properties dialog box.
3. Click Quantities, then Graduated colors.
4. Select the APPRAISAL field for the Value (Figure 135 - 1).

Click the "Add Data" button located in the standard toolbar.

In the "Add Data" dialog, navigate to the "...\ArcBasics\Symbol\SHP" folder and add the "Parcels.shp".

Click OK to close the dialog.

Notice the parcels shape file loads with the default colors.

Click the "Add Data" button located in the standard toolbar.

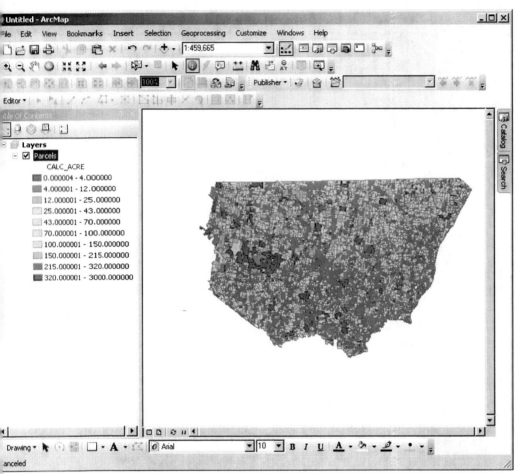

Figure 136: Parcel Layer File

. In the "Add Data" dialog, navigate to the "...\ArcBasics\Symbol\SHP" folder and add the "Parcels.lyr".

Click OK to close the dialog.

Notice how the parcels load in symbolized by CALC_ACRES value (Figure 136).

. Click the "Add Data" button located in the standard toolbar.

. In the "Add Data" dialog, navigate to the "...\ArcBasics\Symbol\SHP" folder and add the "Streets.SHP".

. Click OK to close the dialog.

. Right-click on the streets layer in the TOC and change the street symbology.

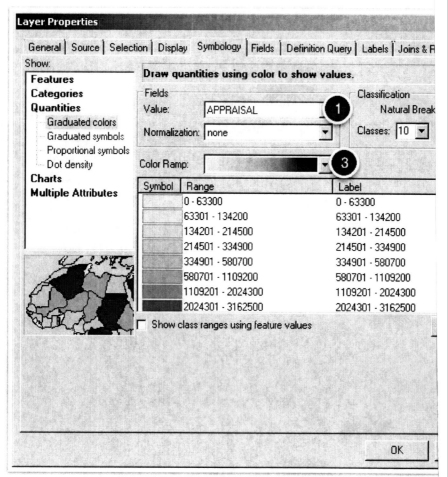

Figure 135: Symbolize by Quantities

5. Change the Classes to 10 (Figure 135 −2).
6. Adjust the color ramp to a scheme you like (Figure 135 − 3).
7. Click "Apply" and "OK". Notice Appraisal value ranges now class
8. Save the Map.

Take a look at See Classifying numerical fields for graduated symb
Help for a good overview on classification methods available in ArcGIS.

#

Create a Layer File (.lyr)

In ArcMap, layer files can be used to save the symbology of a map
symbology is useful in situations where you want to reproduce a map or ha
for your data. Additionally, we have used layer files to allow multiple GIS
with the same look and symbology.

1. Open a new instance of ArcMap.

14. Once the symbology is changed, Right-click on the layer in the TOC and click "Save as Layer File".
15. You have now saved the symbology of the streets layer that can be reused in future maps.
16. Close ArcMap.

#

Data Symbolization Exercise

Use the Symbol MXD to answer the following questions:

1. Use the CALC_Acres field to classify the Parcels layer into 10 classes. What is the low and high range?
2. Add the "Topographic" basemap ("Add Basemap"). Move the topographic basemap below the world imagery in the TOC. Change the transparency of the imagery. At what level do the imagery and the topo meld to form a useful map?

Chapter 7 - Making A Layout

Making a Layout

One of the key functions in ArcMap is the creation of professional maps. ArcMap as a suite of tools that allows you to create a wide variety of maps. These maps can be either traditional hardcopies or digital representations. In this chapter, you will learn the basic for creating maps in ArcMap.

#

Module Objectives:

In this Section, you will learn about:

- The basics of map design
- Symbolizing features and Layers
- Creating layouts
- Configuring the Page Setup
- Adding data frames and graphics to a layout
- Creating a map inset
- Creating and Using Map Templates
- Printing a layout
- Exporting a layout

#

Real World Application

ArcMap has powerful capabilities to create maps. In order to leverage these capabilities, you need to become familiar with some of the basic concepts for map design. The following is the basis of map design.

Before compiling a map, you should answer the following questions:

- ☐ What is the purpose of the map?
- ☐ Who is the audience?
- ☐ What information do you wish your audience to take away after viewing your map?

Determining the answers to these questions is necessary to develop a usable and powerful map.

As with any project, take the time to see how others have addressed a problem. There are numerous examples both online and from other sources of various map styles. Take the time to review how others have answered similar questions as those above with the map. Use the knowledge of others to develop your own mapping style.

#

Map Design

In general, the maps you create should contain the following:

- ☐ Title
- ☐ Legend
- ☐ Scale bar or Scale text
- ☐ North arrow
- ☐ Creation date
- ☐ Creator

Each of these items needs to be placed carefully on the map. Care should be taken so that the map is not cluttered. It is imperative that the items on the map do not detract from the information you wish to convey.

In some cases, it may be appropriate to include a locator map for the area being mapped.

#

Symbolizing Features

When data is added to ArcMap, it is symbolized using a random symbol set. When compiling data to create a map layout, it is often important to change the default symbol set. ArcMap contains thousands of possible symbols and colors that can be used to make a more informative map. In the following exercise you will learn how to symbolize data.

1. Open a new instance of ArcMap.
2. Open the "Layout.mxd" file located in the "...\ArcBasics\Layout" folder.
3. In the Robertson County map, you will see nine different layers. You will be using these layers to create a map for the city of Springfield.

It is important that you ask three questions before creating the maps. In the case of the city of Springfield, the map will be for the following purpose:

What is the purpose of the map?

☐   The purpose of this map will be to show the commercial properties in Springfield.

Who is the audience?

☐   The map will allow a developer to look at the commercial properties located in central Springfield.

What information do you wish your audience to take away after viewing your map?

☐   It is critical that the map show utilities, fire hydrants, zoning, streets, and parcels. The audience must be able to determine the commercial properties and their attributes.

1.   In the TOC, turn on the "Springfield" layer. Change the city limits to be hollow and show the border in a better color. Zoom to the layer extends.
2.   Change the symbology of the "Parcels" layer and symbolize the parcels by property type.
3.   Change the "WaterBodies" layer to be symbolized appropriately (Figure 137).

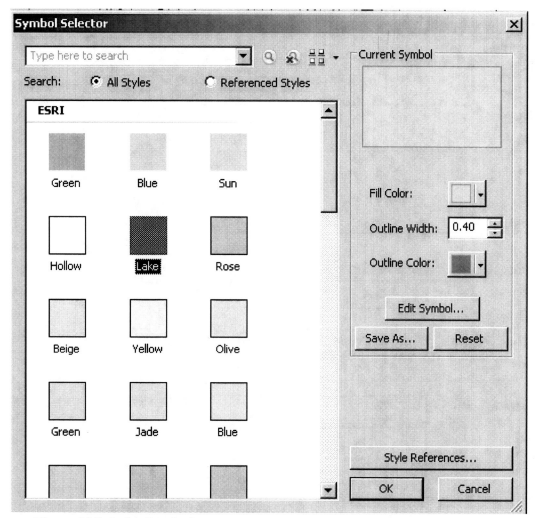

Figure 137: Lake Symbols

4. For the "Schools" layer, select as a school image.
5. Symbolize the various utility layers appropriately.
6. Turn on the "Streets", turn the labels on for the streets, and use street name field.
7. Once you are satisfied with the appearance of the map, save the map to the "...\ArcBasics\Layout" folder.

#

Set a Reference Scale for a Data Frame

Using the map you created in the previous section, you will set a reference scale for the data frame. Setting a reference scale will allow you to use a larger data set but set the map to only use at the fine scale. For example, you can have a data set of Robertson County but you can set the map to only show the city of Springfield. In the following exercise, you will use a reference scale of 1:72,000.

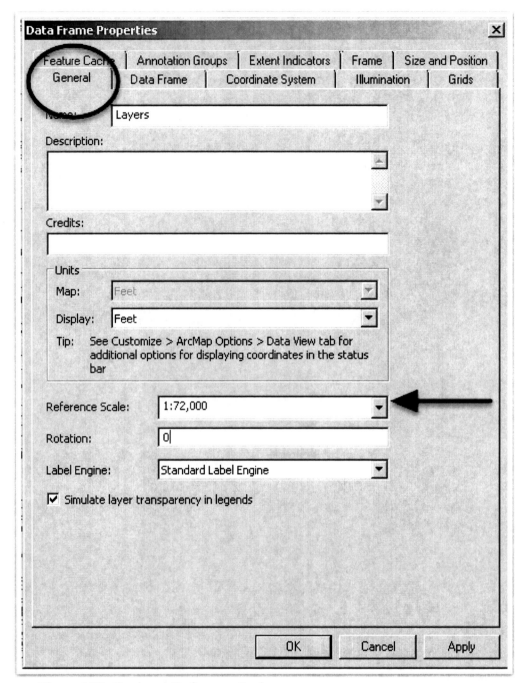

Figure 138: Setting a Reference Scale

1.   Right-click on the "Layers" Data Frame.
2.   Click on the "Properties" in the context menu.
3.   Click on the "General" tab in the "Data Frame Properties" dialog (Figure 138).
4.   Type "72,000" in the "Reference Scale" text box.
5.   Click "OK" and the map display will automatically zoom to the set scale.
6.   Using the "Zoom in" tool in the standard toolbar, zoom in to the central part of the map.

7.   Right-click on the data frame in the TOC, click on "Reference Scale" and click on "Zoom to Reference Scale".

8.   The map has now zoomed back to the reference scale.

#

Preparations for Creating a Layout

When creating the layout in ArcMap, you need to use " Layout View". The layout view is set up to mimic a sheet of paper. This virtual Canvas allows you to create, add, and edit items on the map. These items include titles, scale bars, scale text, legends, etc.

In the following example, you will begin to explore preparing a map in the layout view. Initial preparations include turning on the various layers needed and zooming the map to an appropriate scale. As stated previously, we will create a map of the commercial properties in Springfield.

1.   Open ArcMap and load the "Layout" MXD file located in the "...\ArcBasics\Layout" folder.

2.   Turn on the following layers: parcels, streets, schools, water bodies, and city limits (Figure 139).

3.   Zoom to the central portion of the city.

4.   Change the order of the layers in the TOC.

5.   Make sure the symbology is appropriate.

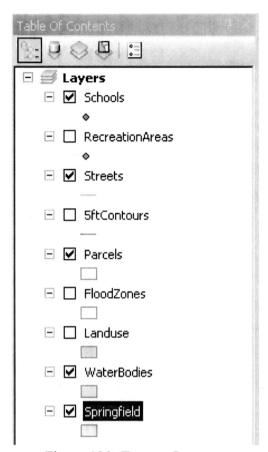

Figure 139: Turn on Layers

#

Switch to Layout View

In ArcMap, you use the "Layout View" to construct a map.

There are two methods to access the layout view:

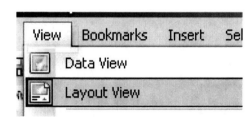

Figure 140: View Layout

1.   Click "View" in the standard menu bar and then click "Layout View" (Figure 140).

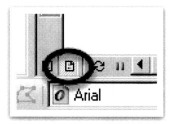

Figure 141: Layout Button

2.   Use the "Layout View" button located in the lower left of the view window (Figure 141).

When you enter the "Layout View", the "Layout" toolbar becomes active (Figure 39). The "Layout" toolbar has a variety of tools to interact with the map layout. Although the tools look similar to those in the "Standard" toolbar, there is a difference. The "Standard" tools, such as "Zoom In", interact with the data regardless of the view. For example, the standard "Zoom In" will cause the map to change scale in all views. When using the "Layout Zoom", only your perspective on the layout changes. Notice that when you change back to the "Data View", the map is still at the original scale.

Figure 39: Layout Toolbar

#

Set the Layout Page Size and Orientation

When creating a map, it is important to consider the page size for the layout. It is advisable to set the page size before beginning to place elements on the map. If you do not and you change the map layout, the elements will need to be moved and/or resized. In the following exercise, you will learn how to set both the size and orientation of the layout.

Figure 142: File -> Page and Print Setup

1. To change the paper size and layout, click on "File" in the menu and click "Page and Print Setup" (Figure 142).

2. Use the "Orientation" radio buttons to change the layout from portrait to landscape (Figure 143).

3. Make sure the "Size" drop-down is set to "US Letter".

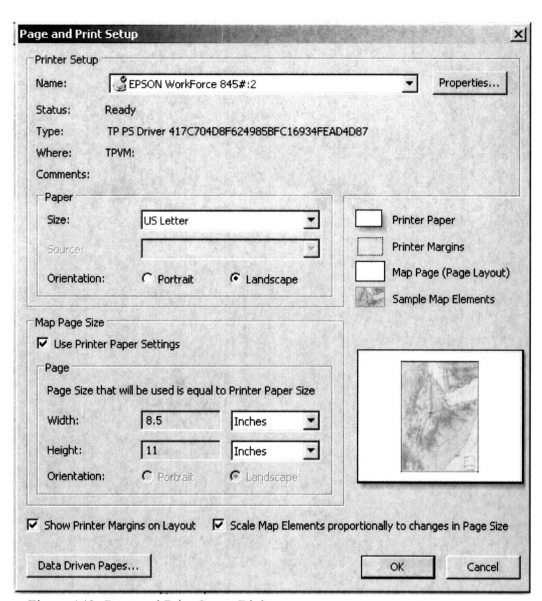

Figure 143: Page and Print Setup Dialog

4. Check the "Use Printer Page Settings" checkbox. This Setting will force the width and height settings to match the printer. There are cases when you may want these settings to change.

5. Check the "Scale Map Elements Proportionately To Changes in Paper Size" checkbox. Using this setting will automatically scale map elements. Keep in mind that this setting will not in all cases perfectly scale map items. You should be sure to check all the elements to verify appropriate sizes.

6. Click "OK".

#

Move and Resize a Data Frame in a Layout

1. Click on the "Select Elements" tool from the Tools toolbar.
2. Click on the data frame in the layout. The frame is highlighted and has eight square "Handles". Use the handles to resize the map (Figure 144).

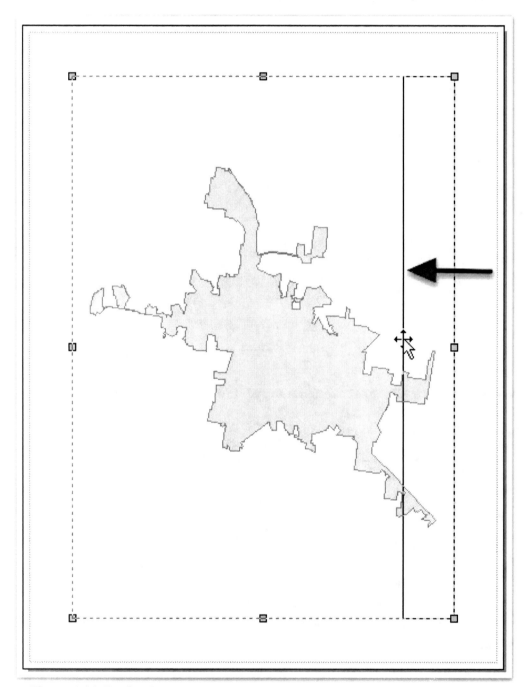

Figure 144: Resize Layout

3. To move the map on the layout, click and drag the map to a new location, and then release the mouse button.
4. Move the map to the left side of the layout and resize it so that additional elements can be placed above and on the right.

#

Manipulate Graphic Elements

As you create a map layout, you will be adding various graphic components, such as a North arrow, a legend, images, etc. Once these graphic elements are added, you may need to move and change the graphics. Many graphic commands are available by Right-clicking on a graphic. The graphic context menu (Figure 145) contains these options:

Figure 145: Graphic Context Menu

- Selection: the "Select Elements" tool can be used to select graphics.
- Unselect: To unselect a graphic, you can click on the area outside of the layout.
- Deletion: To delete a graphic, select the graphic and press the delete key.
- Move: A graphic can be moved by selecting the graphic, and dragging the image to a new location.

- Resizing: The size of the graphic can be changed by selecting the graphic and moving the cursor to the graphic corner and dragging to the desired size.
- Grouping: Multiple graphics can be grouped together, by selecting the desired graphics, Right-clicking, and clicking on "group" in the context menu.

At any point, you can undo an action by going to the "Edit" in the menu bar and clicking "Undo" (Ctrl-Z).

#

Insert a Title

Your maps may require a title. In this exercise, you will learn how to add a title to a map by using the drawing toolbar and the insert menu.

Drawing Toolbar

Figure 146: Add a Title via the Drawing Toolbar

1. Click the Arrow next to the "Text/label" tool on the "Drawing" toolbar.
2. Click on the "Text (A)" tool.
3. Click on the location where you would like to place the title.
4. Press the "Enter" key.
5. Make sure the text is selected, and change the font size, as needed, in the "Drawing" toolbar.
6. Double-click on the text to open the "Text Properties" dialog.
7. Type in the appropriate title.
8. Click "OK".

Insert Menu

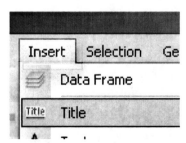

Figure 147: Add a Title via the Insert Menu

1. Click on the "Insert" menu and click "Title" (Figure 147).
2. Double-click on the title text box.

3. Double-click on the text to open the "Properties" dialog.
4. Type in the appropriate title.
5. Click the "Change Symbol" button, and change the font size and style, as needed.
6. Click "OK".
7. Drag the title to the desired location.

#

Insert a Legend

To add a legend, use the following steps:

1. Click on "Insert" from the menu, and then click "Legend".
2. The "Legend Wizard" will open (Figure 148). The box to the left shows the available layers, the box to the right is the layers in the map. You can move layers on and off of the legend as needed.
3. Click "Next" proceed through the wizard.
4. Click "Finish".
5. You can drag the legend to a new location or resize.

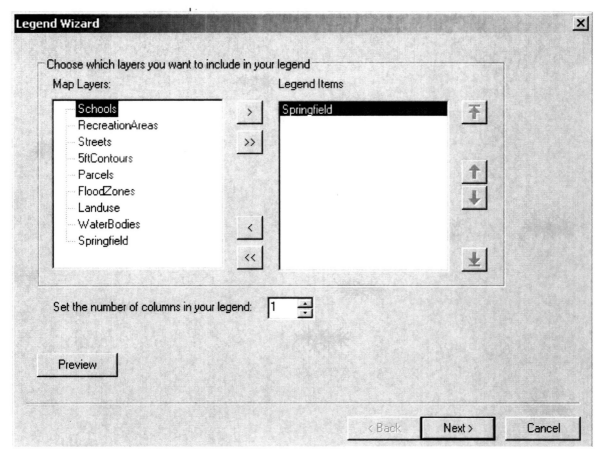

Figure 148: The Legend Wizard

You can turn layers on or off in the TOC. Layers that are on will show in the legend. Those that are turned off will not be present.

Even though you have finished the legend wizard, you can double click on the legend to change properties. Some of the most common changes you will make will be to add a background to the legend or add an outline.

Note: In some cases, you may want to make some very different changes to the legend. It is possible to convert the legend to graphics and edit individual lines of text and color swatches. Take care when converting a legend graphic, once a legend is graphic it is no longer dynamic and cannot be converted back.

#

Insert a North Arrow

1. Click on "Insert" in the menu and then click "North Arrow". The "North Arrow Selector" dialog will open (Figure 149).
2. There are a variety of North arrows available. Review the north arrows and click on the one most appropriate.

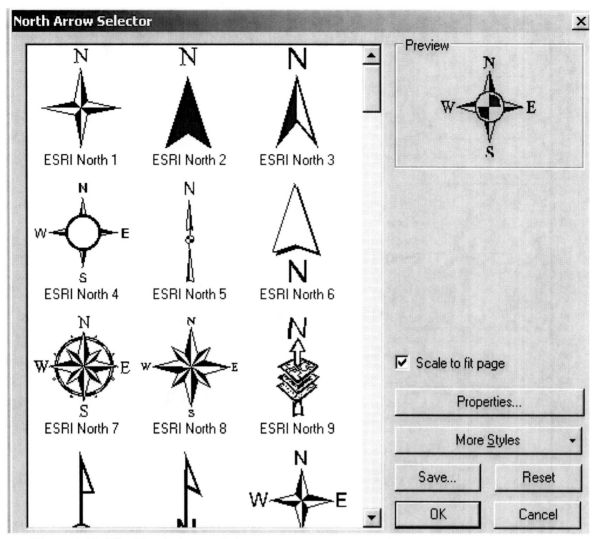

Figure 149: The North Arrow Selector

3. Click "OK" to add the selected north arrow to the layout.
4. With the North are selected, drag the north to an appropriate location. Use the graphic handles to size it appropriately.
5. Click outside the layout to deselect the north arrow.

#

Insert a Scale Bar

When presenting a map at a known scale, it is important to include a scale bar on the map. In the following exercise, we will add a scale bar to the layout.

1. Click the "Insert" menu, and click "Scale Bar". The "Scale Bar Selector" dialog will open and show a variety of available scale bar styles (Figure 150).

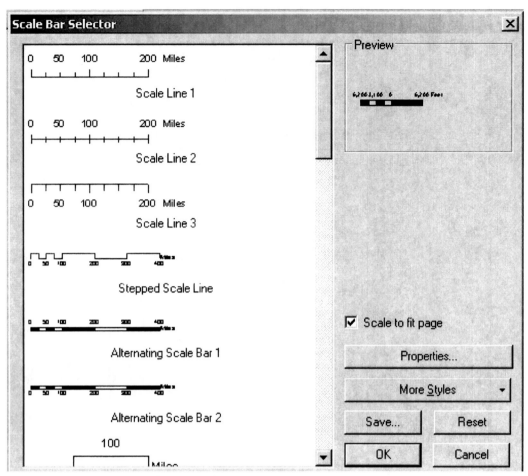

Figure 150: The Scale Bar Selector

2. Click on the "Scale Line 1" and click on the "Properties" button on the dialog window (Figure 151).

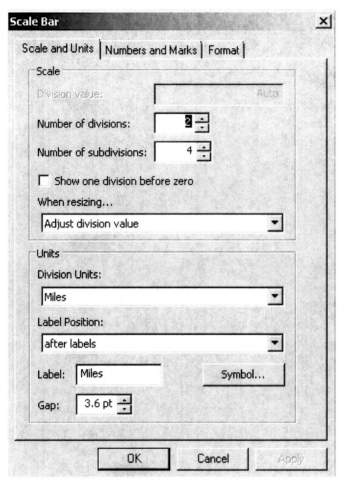

Figure 151: The Scale Bar Properties

3. When the "Scale Bar" dialog opens, click on the "Scale and units" tab and set the "Number of Divisions" number spinner to 4.

4. Set the "Number of Subdivisions" number spinner to 4.

5. Check the "Show one division before zero" box.

6. Set the "Division Units" drop-down to the Feet.

7. Click "OK" in both dialogs to place the scale bar on the map.

8. Drag the scale bar to an appropriate location.

9. Double-click on the scale bar to access the scale bar properties as needed.

10. Click outside the map layout to deselect the scale bar.

#

Insert Text

When creating a map, you may need to insert text onto the map. I would strongly recommend that on each of your maps you insert a text box containing your name, map date, and the name of the MXD used. This type of information can be critical if you need to reproduce a map at a letter date. In the following exercise, you will learn how to insert a text box containing this information.

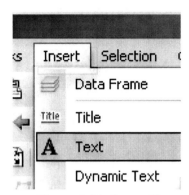

Figure 152: Insert Text

1. Click on "Insert" in the main menu, and then click "Text" (Figure 152). A text box will be inserted onto the map.
2. Insert the text: "<<Your Name>> ,<<Date>>, <<MXD Name>>" into the text box and press the "Enter" key.
3. In the "Drawing" toolbar, change the text size drop-down to 12 or an appropriate size.

1. Drag the text box to an appropriate location.

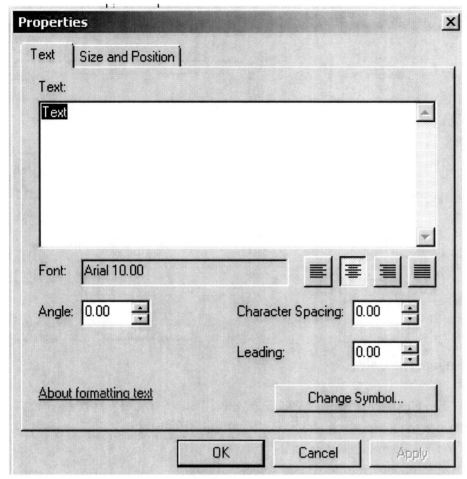

Figure 153: Text Properties

4. Double-click on the text box to access the Text properties as needed (Figure 153).
5. Click outside the map layout to deselect the text box.

#

Insert a Neatline

A Neatline is an outline or box that frames the map and gives it a finished look.

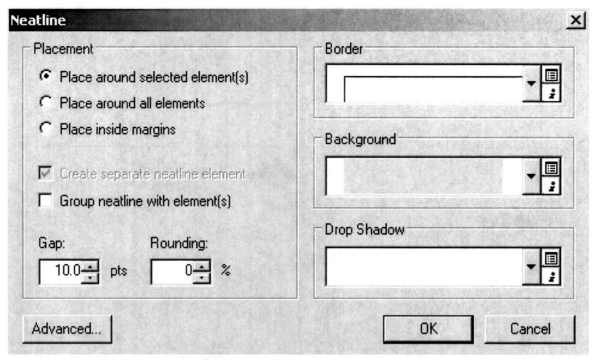

Figure 154: Neatline Properties

1. Select the Map frame using the "Select Element" tool.
2. Select "Insert" from the main menu and click "Neatline" (Figure 154).
3. Change the "Gap" number spinner to 4 pts.
4. Click on the "Background" drop-down and select "Hollow".
5. Click the "OK" button.

#

Align Graphic Elements

In order to create a professional looking map, it is important that map elements are aligned and symmetrical. With ArcMap, you can align graphic elements in several different ways. For example, you can center the scale bar under the data frame by taking the following steps:

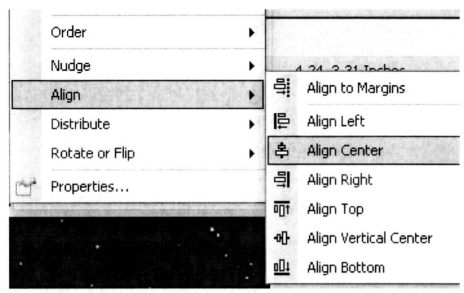

Figure 155: Align Center

1. Click on the "Select Elements" tool, click on the scale bar, and while holding down the <Shift> key and also select the data frame.
2. With both elements now selected, right-click on the data frame and click "Align". Then click "Align Center" (Figure 155). This will center the scale bar in under the data frame.

#

Create a Map Inset

Adding a map inset to your map is a good way to orient the map user for the map location. In the following exercise, you will learn how to place an inset.

1. Click on "Insert" in the Main Menu, and click "Data Frame". A new, empty data frame will be added to the TOC.
2. Right-click on the new data frame later and click "Add Data".
3. Browse to the inset data located in the "...\ArcBasics\Layout" folder. Add the County layer to the map.
4. In the TOC, right-click on the new data frame and click properties. The "Data Frame Properties" dialog will open.

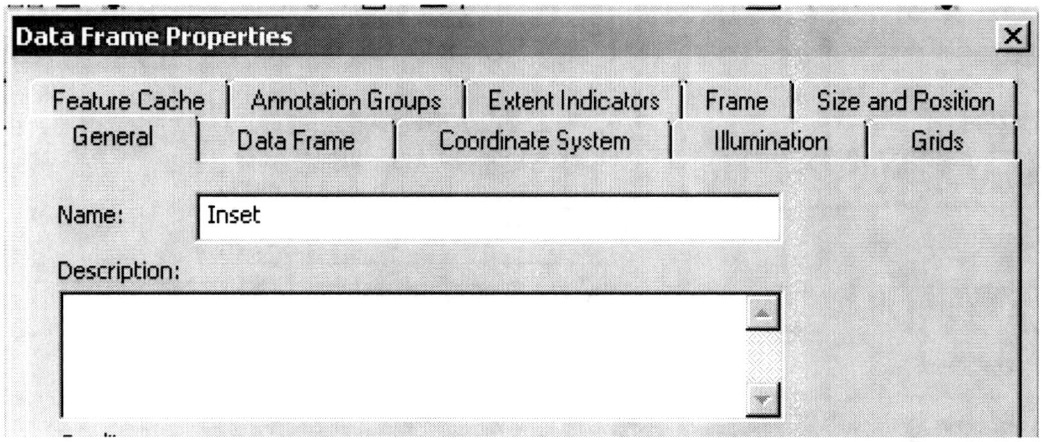

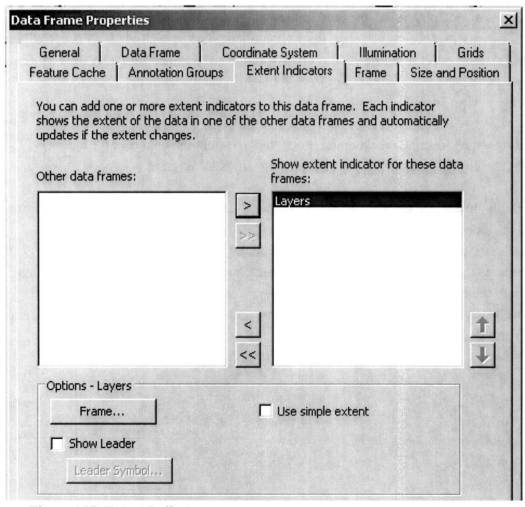

Figure 156: Change Inset Name

5. Select the "General" tab, and change the "Layer Name" to Inset (Figure 156).
6. In the "Extent Indicators" tab, select the "Data Frame" name and add it to the "Show extent indicator..." list (Figure 157).

Figure 157: Extent Indicator

7. Click "OK" twice to close the dialog.

8. Use the "Select Elements" tool to move and size the inset appropriately.

9. Use the "Data Zoom" tool, to set the ideal scale in the inset.

10. Click outside of the layout to deselect all elements.

#

Add Graphics to a Data Frame from the Layout View

Similar to the way you add graphics to the data view, you can add graphic elements (such as text to a data frame from the layout view). Graphics added to the layout view will not be added to the data view. Additionally, these graphics will not be tied to a specific location. In this section you will add text to the inset data frame from the layout.

1. Double-click on the "Inset" data frame in the layout to give it focus.

2. Use the "Zoom In" tool on the "Layout" toolbar and zoom in on the "Inset".

3. Select the "New Text" tool from the "Draw" toolbar and click to the right of the extent rectangle, type "Area of Interest", and hit <Enter>.

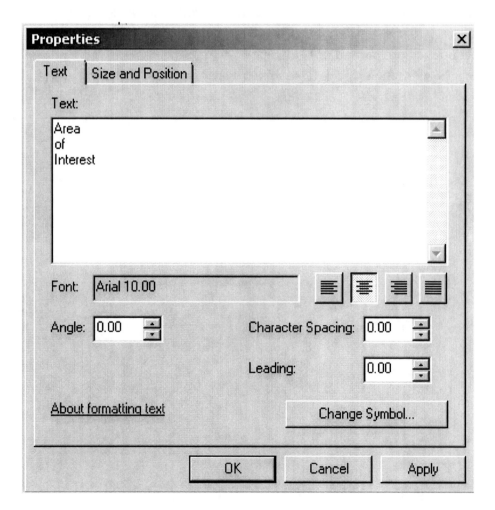

Figure 158: Insert Text in Layout View

4. Double-click on the text, "Area of Interest" and in the "Properties" dialog in the "Text:" textbox and place the cursor after "Area" and hit <Enter> key; repeat the process to create 3 lines of text Figure 158).
5. Click on the "Center Justification" button to center the text.
6. Click on "Change Symbol...." and select Garamond, Size 36, and Style Bold.
7. Click "OK" to close the dialog."
8. Use the "Select Elements" tool to move the "Area of Interest text to place it better in the "Inset" data frame.

#

Print a Layout

1. Click "File" in the main menu, and then click "Page and Print Setup" (Figure 159).
2. In the "Print and Page Setup" dialog, click on "Properties" and choose the large format plotter.
3. Set the paper size. We will use Arch D, which is 24 in x 36 in. I would recommend consistently using "Architectural" type paper sizes. Engineers and surveyors typically use these sizes.
4. Select the paper orientation (portrait or landscape).
5. In the "Map Page Size" section, check "Use Printer Paper Settings" checkbox.

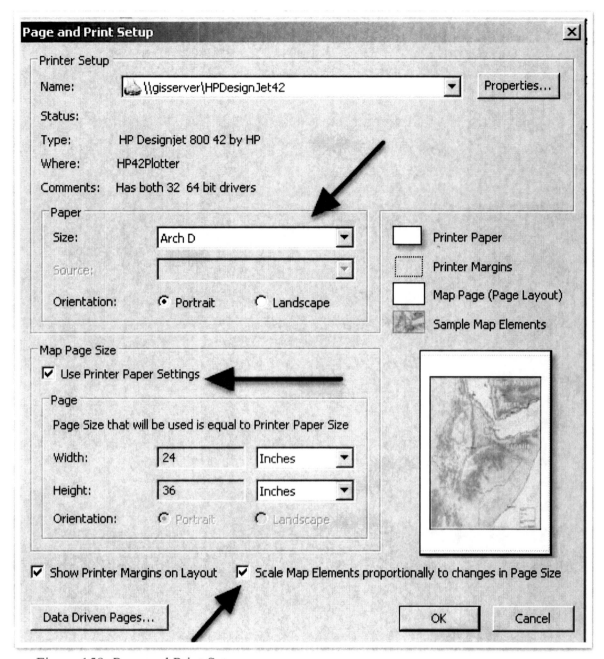

Figure 159: Page and Print Setup

1. Check the "Scale Map Elements Proportionately To Changes in Paper Size" checkbox. Using this setting will automatically scale map elements. Keep in mind that this setting will not in all cases perfectly scale map items. You should be sure to check all the elements to verify appropriate sizes.

2. Check the sample map illustration on the right. Make sure that your map area covers the entire paper and is aligned properly.

3. Click "OK" to close the "Print and Page Setup" dialog.

4. Always save your map prior to Printing (Step 10).
5. Click "File" in the main menu, and then click "Print". Your document will be sent to the plotter.

#

Export a Map

Once you have created a map you may want to export it to another format. For example, you may want to email a copy or use the map in a presentation.

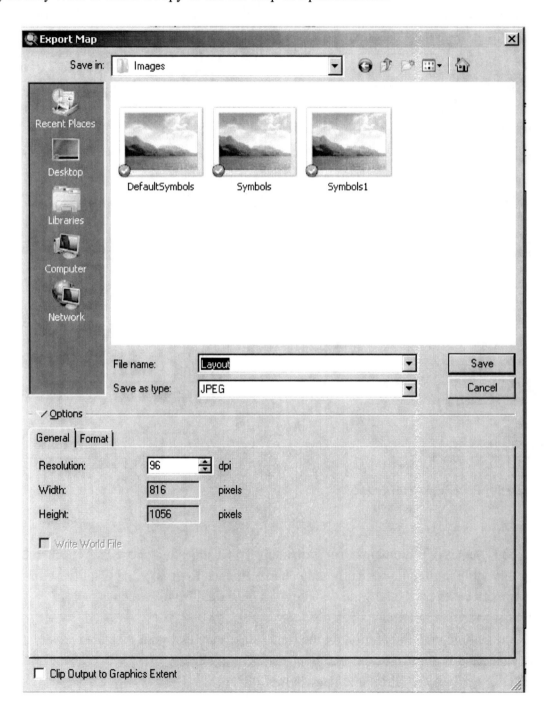

Figure 160: Export Map

1. Click the File menu and Choose Export Map....
2. Once the Export Map dialog opens (Figure 160), click on the dropdown to the right of the Save in: window and navigate to the folder where you want to save the export.
3. Click the "Save as type" dropdown and select an export format (PNG, PDF, BMP, TIFF, JPEG, etc.). Two of the most commonly used formats include PDF and JPEG.
4. In the File name Textbox, type a file name.
5. In some cases, an "Options" button will be available. This button will give you access to a variety of export format options. Some example options include image quality, background color, PDF options, etc. Feel free to experiment with the available options.
6. Click Save.

Chapter 8 - Mid-Term Exercise 1

The Problem

Note: Read through the entire problem before starting.

The City of Springfield requires two maps in order to review the growth plan. The two maps needed are:

1. A map displaying the properties surrounding the City classified by property type.
2. A flood map is needed with the flood zones and roads displayed.

To complete this project, you will be required to load the data layers in "...\ArcBasics\Mid-Term1\SHP".

The layers for the Growth plan you need to include:

1. County.shp
2. Springfield.shp
3. Parcels.shp
4. Streets.shp
5. Waterbodies.shp

For the Growth plan map, you will need to add the following items:

- Map title
- Legend
- North Arrow
- Scale bar
- Streets labels
- Property Types which are classified and labeled
- A Text Box with a Map date and Prepared By name

The layers for the Flood Map should include:

1. County.shp
2. Springfield.shp
3. FloodZones.shp

4.   Streets.shp
5.   Waterbodies.shp

For the Flood map, you will need to add the following items:

- Map title
- Legend
- North Arrow
- Scale bar
- Street labels
- Flood Data that is classified and labeled
- A Text Box with a Map date and Prepared By name

Chapter 9 - Using Attribute Data

Using Attribute Data

The GIS allows the user to store, visualize, and analyze database information. The foundation of all GIS layers is each layer's attribute data. In order to fully realize the power of GIS, you must become familiar with displaying, using, and manipulating attribute data.

#

Objectives

At the conclusion of this section, you will be able to:

☐ Display and manage attribute tables
☐ View statistics for a field
☐ Summarize a field
☐ Delete an existing field
☐ Add a new field
☐ Calculate values for a field
☐ Join and relate tables
☐ Establish hyperlinks

#

Real World Application

Hopefully you are starting to realize that much of the work in GIS is similar to work done by a database administrator. So far, you have learned how to display spatial data and to utilize query functionality. What happens when you notice that something in the attribute table is spelled wrong or otherwise incorrect? As a GIS practitioner, you will need to know how to manipulate and edit attribute table. This section will show you the basics of working with attributes.

When I start to discuss editing attributes, many people immediately start to become concerned with "messing up their data". Mistakes happen to everyone and are nothing to be feared. If you are worried about making a mistake, use the SAVE button.

Lastly, keep in mind that manipulating GIS attributes are very similar to those you are probably using regularly to work in Excel.

#

Components of an Attribute Table

As you have already seen, an Attribute Table contains descriptive information about a data layer. An attribute table is organized like a simple spreadsheet or database. Each row of information is called a record, each column of data is called a field, and each piece of information is called a cell. ArcGIS 10.1 or 10.2 supports several sources of attribute data including, dBase files, text files, and geodatabase tables.

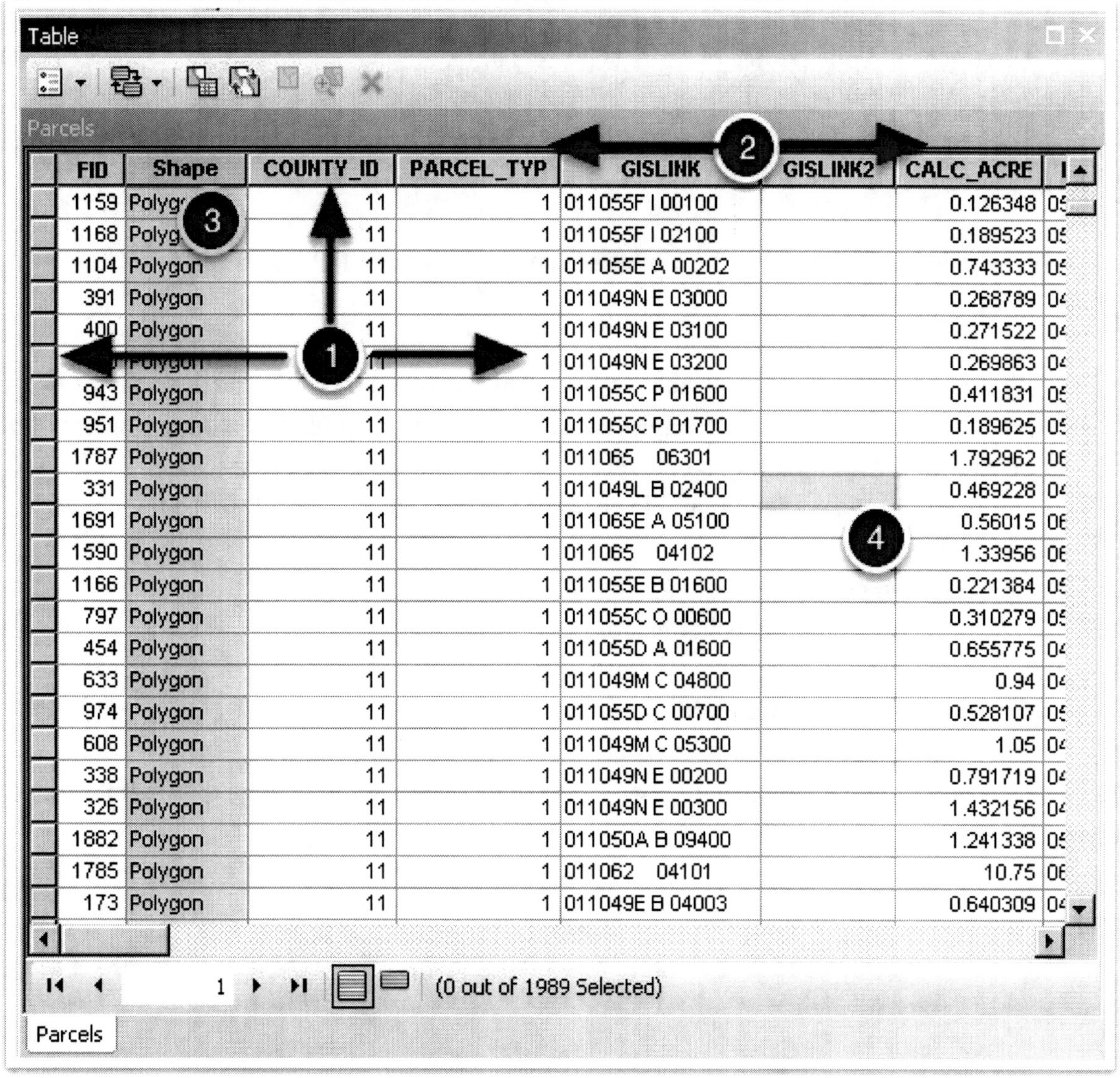

FID	Shape	COUNTY_ID	PARCEL_TYP	GISLINK	GISLINK2	CALC_ACRE
1159	Polygon	11	1	011055F I 00100		0.126348
1168	Polygon	11	1	011055F I 02100		0.189523
1104	Polygon	11	1	011055E A 00202		0.743333
391	Polygon	11	1	011049N E 03000		0.268789
400	Polygon	11	1	011049N E 03100		0.271522
[illegible]	Polygon	11	1	011049N E 03200		0.269863
943	Polygon	11	1	011055C P 01600		0.411831
951	Polygon	11	1	011055C P 01700		0.189625
1787	Polygon	11	1	011065 06301		1.792962
331	Polygon	11	1	011049L B 02400		0.469228
1691	Polygon	11	1	011065E A 05100		0.56015
1590	Polygon	11	1	011065 04102		1.33956
1166	Polygon	11	1	011055E B 01600		0.221384
797	Polygon	11	1	011055C O 00600		0.310279
454	Polygon	11	1	011055D A 01600		0.655775
633	Polygon	11	1	011049M C 04800		0.94
974	Polygon	11	1	011055D C 00700		0.528107
608	Polygon	11	1	011049M C 05300		1.05
338	Polygon	11	1	011049N E 00200		0.791719
326	Polygon	11	1	011049N E 00300		1.432156
1882	Polygon	11	1	011050A B 09400		1.241338
1785	Polygon	11	1	011062 04101		10.75
173	Polygon	11	1	011049E B 04003		0.640309

Figure 161: Attribute Table

You have used attribute tables in previous sections and have discussed elements of the attribute table. Using Figure 161, let's take a look at the key components of a table. Right-click on the "Parcels" layer in the TOC. The attribute table for the parcel layer will open.

1)   As you've already seen, the attribute table is similar to a spreadsheet. The table is divided into rows and columns. Each column is known as a field. When the parcels attribute table opens notice the various fields.

2)   Notice the field headings have a different background than the cells. The different heading color indicates that the heading is not editable.

3)   Each row in the attribute table represents one record, each record being a spatial feature.

4)   Each block in the table is known as a cell.

#

Change the Appearance of an Attribute Table

There are several different ways in which you can change the appearance of an attribute table. For example:

1.   Records can be sorted in the table based upon a selected field or fields (see Chapter 4 - Displaying Spatial Data).

2.   Change the width of fields.

3.   Freeze a field so that it is visible as you scroll horizontally in the attribute table.

4.   Hide a field so that it is not displayed. It is important to note, that hidden fields will not be exported when creating a new record set based upon a selection.

5.   You can create an alias for a field. Aliases are particularly useful to display a more descriptive field name.

6.   The user can change the color of the selected row.

7.   The ordering of the fields in the table can be changed.

#

Change the Width of a Field

Depending on the data being viewed, it may be necessary to change the width of the field in the table.

1.   Open the "UAD.mxd" map document located in the "...\ArcBasics\UAD" folder.

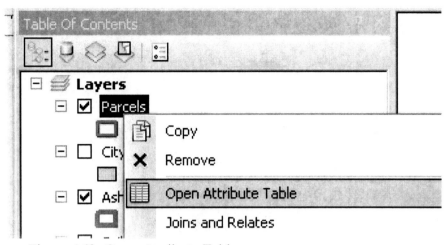

Figure 162: Open Attribute Table

2. Right-click on the Parcels layer name in the TOC and click Open Attribute Table (Figure 162).

3. The Attributes of Parcels table opens. Position the pointer to the right of the FID field heading and between the FID and Shape* field headings; the pointer symbol changes to a bidirectional arrow (Figure 163).

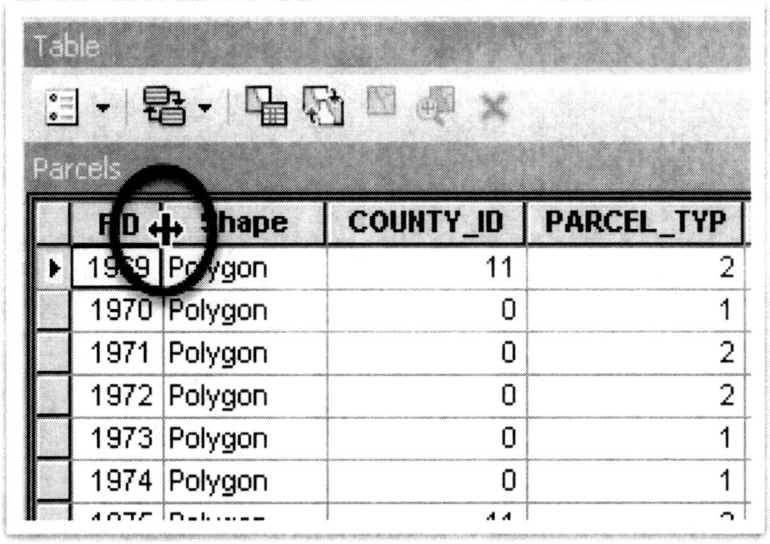

Figure 163: Bidirectional Arrow

4. Click and drag to the left to decrease the field width. To increase a field width click and drag to the right (Figure 164).

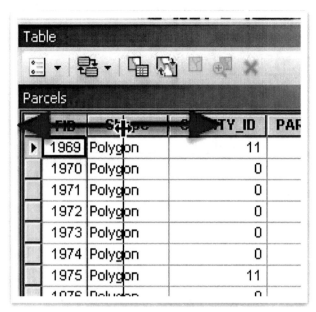

Figure 164: Drag to Change Column

5. Adjust the widths of other fields in the table.
6. Close the Attribute table when finished.

#

Freeze Fields in an Attribute Table

When modifying attribute tables with numerous fields, it is often advantageous to use the freeze field function in ArcMap. When you freeze a field, the field is always visible as you scroll horizontally. The frozen field stays visible on the left-hand side of the table regardless of how far you scroll. Frozen fields are separated from non-frozen fields by a thick black line. To freeze a field use the following steps:

1. Open the UAD.mxd located in the "...\ArcBasics\UAD" folder.
2. In the TOC, right-click on the parcels layer and click on "Attributes of Parcels". The attribute table will open.
3. Find the "GISLINK" field in the table and right-click on the field heading.
4. In the context menu, click the "Freeze/unfreeze column" option (Figure 165).

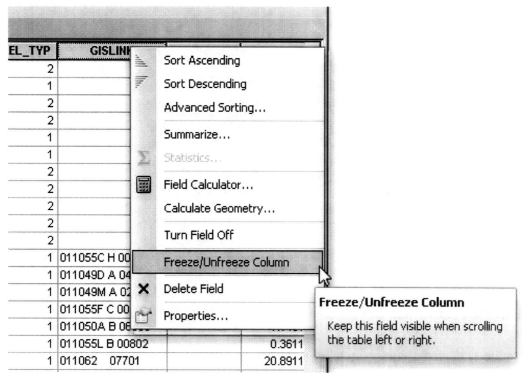

Figure 165: Freeze Column

5. Scroll to the left of the table and notice how the "GISLINK" field stays visible. Find the "OWNER" field, and freeze the column. Notice that both fields are always visible on the left-hand side of the table. Additionally, the fields are separated from the non-frozen fields by a thick black line (Figure 166).

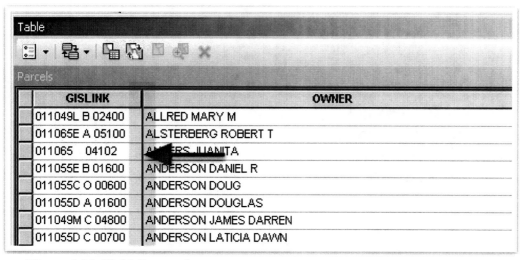

Figure 166: Field Separation

6. Right-click on the "GISLINK" field and unfreeze the field. Now only the "OWNER" field stays in place.

7. Unfreeze the "GISLINK" field.

#

Hide Fields in an Attribute Table

When working with spatial data, the attribute table may contain a large number of fields that are irrelevant for your particular study. In these situations, ArcMap has tools that allow you to hide fields in an action be table. Hiding fields helps you to avoid constantly have to into scroll through large tables and irrelevant fields. It is important to note that although the field may be hidden it is still present in the data set. Hiding the field does not delete it. When using selections to create new data sets care should be given when fields are hidden. In some cases, hidden fields will not export. In the following exercises, you will learn how to hide a field.

Open the UAD.mxd located in the "...\ArcBasics\UAD" folder.

Via the Attribute Table:

1. From the TOC, right-click on a layer and open its attribute table.
2. Use the scrollbar to look through the data set.

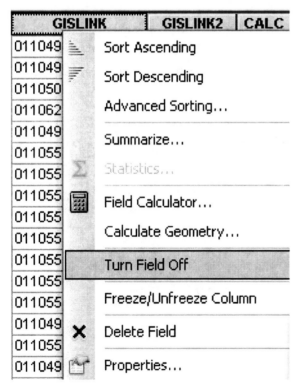

Figure 167: Turn Field Off via the Attribute Table

3. Right-click on a field heading and click on "Turn Field Off" (Figure 167). The selected field is now hidden.

Via the Layer Properties:

1. With the attribute table open, use the scrollbars to browse the table.
2. Right-click on the "Parcels" player in the TOC and select "Properties".
3. Click on the "Fields" tab in the properties window and click the check mark next to the "COUNTY_ID" field from the list. By unchecking the field, you have hidden it from view (Figure 168).
4. Click Apply.
5. Notice that the "COUNTY_ID" field is no longer visible.
6. Repeat the process with fields you select to test the functionality. Click Apply when finished and review the results.

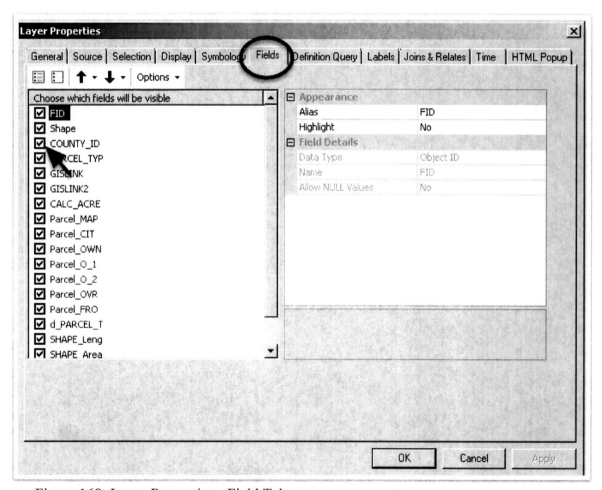

Figure 168: Layer Properties - Field Tab

#

Create an Alias for a Field Name

In many cases, field names and attribute table are not descriptive. You can use aliases to give the fields more descriptive names. For example, an alias can contain characters like spaces and symbols. Actual field names cannot contain these items and must be letters or numbers.

1. Open the "Parcels" layer attribute table. Notice the "Parcel_MAP" field.
2. Open the "Layer Properties" dialog for the "Parcels" layer.
3. Click on the "Fields" tab in the dialog.
4. Select the "Parcel_MAP" field from those listed and in the alias column replace "Parcel_MAP" with "Map" (Figure 169).

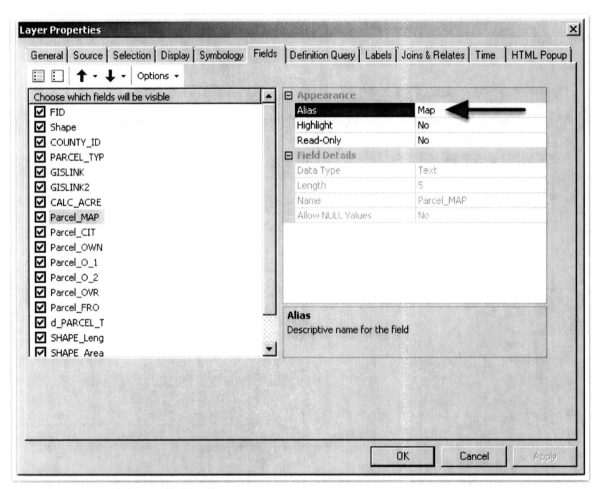

Figure 169: Using a Field Alias

5. Click okay to close the "Layer Properties" dialog and notice that the "Parcel_MAP" field has been renamed "Map" in the attribute table.
6. Close the attribute table.

#

Setting the Highlight Color for a Table

When working with tables, you may want to change the selection color of a selected row. In the following example, you will see how to change the highlight color.

1.   Open the "Parcels" attribute table. In the action view table click on the "Table options" button and click on "Select by attribute".
2.   In the "Select by attribute", in the "Method" combo box click on "Create a new selection". In the "Fields" list box, double-click on "CALC_ACRE", click on the ">" button, and type in "1". The expression box should look like Figure 170.

SELECT * FROM AshlandCityParcel WHERE:

"CALC_ACRE" >1

Figure 170: Select by Attribute

3.   Click "Apply" and then click "Close".
4.   In the "Attributes of parcels" table click on the "Selected" button. Notice that the selected records are light blue.
5.   Use the "Select element" tool and click on the gray box to the left of one of the selected records. By default, the highlighted record is shown in yellow.

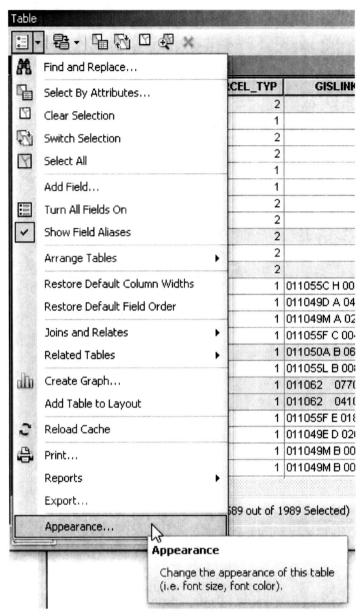

Figure 171: Table Options Appearance

6. To change the highlight color, click on the table "Options" button and click on "Appearance" (Figure 171).

7. In the "Table appearance" dialog, click the drop-down arrow for the option "When table is only showing selected records, use the color for highlighted records and for their features" and pick a different color (Figure 172).

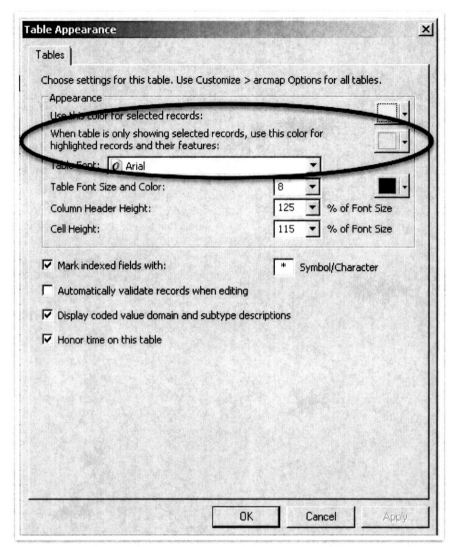

Figure 172: Table Appearance Dialog

8. Click "Apply" and then click "OK" to finalize the new highlight color.
9. Close the Attribute table.

#

Field Properties

As you've probably already noticed, you can open the "Layer Properties" dialog to view a Field's properties.

1. Open the "Layer Properties" dialog for the Parcels layer and click on the "Fields" tab.
2. The layer properties dialog now displays the various field properties. These properties include "Alias", "Highlight", "Read-only status", "Datatype", "Length", "Name", and "Allow null values" for each field in the attribute table. Many of these values are set

when the field is created. The property values that cannot be changed are in light gray, while the editable items are in black (Figure 173).

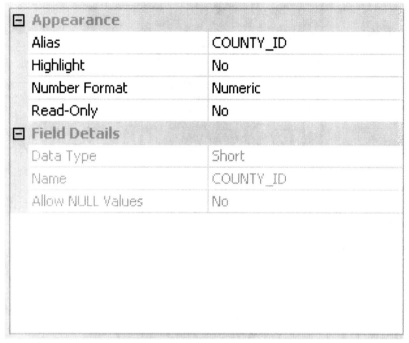

Figure 173: Field Properties

Field Type

There are eight field types available in art GIS. These field types are:

1. "Short and long data types" – used to store numeric values. Often used for numeric values that can be counted or is used to designate classifications.
2. "Float and double data types" – used to store real numeric values such as those that are used for continuous data such as measurements or calculations.
3. "Text fields" – used to store character values such as feature name and descriptions.
4. "Date fields" – used to store temporal data such as time and date.
5. "Blob fields" – used to store binary data such as images and media.
6. "Global identifier fields" (GlobalID and GUID) – used to store registry style strings that are unique identifiers in geo-databases.

Field length

Field length is the maximum number of characters that can be stored in field. For text fields, this is typically 256 characters.

Precision and Scale

Precision is the number of digits that can be stored in a field. The higher the precision, the more disk space that will be required for storage.

Scale is the number of decimal places for fallout and double fields.

ESRI suggests the following guidelines for choosing the correct field type for various levels of precision and scale:

☐ Precision is six or less, use a float field.
☐ Precision is greater than six, use a double field.
☐ Scale of zero and precision of 10 or less, use integer field.

#

View Statistics for a Field in an Attribute Table

You can obtain statistical values for numeric fields in an attribute table.

The obtainable values include count, minimum, maximum, sum, mean, and standard deviation.

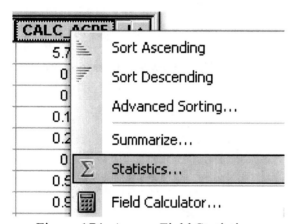

Figure 174: Access Field Statistics

1. In the "Parcels" table, find the "CALC_ACRES" field. Right-click on the field and click on Statistics (Figure 174).
2. The dialog opens. The dialog displays the various statistical values for the Acres field (Figure 175).

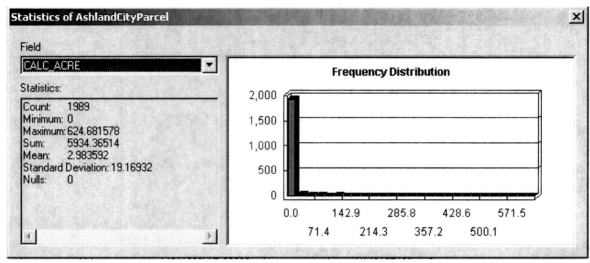

Figure 175: Field Statistics

3. Notice the dialog also shows a "Frequency distribution" chart.

4. Close the statistics dialog.

#

Summarize a Field in an Attribute Table

In ArcMap, the "Summarize" function allows you to create tables that contain the summary statistics for selected field in a natural table. In the folly example you will create a dBase table based on the unique values in "MUSYM" field (Soil Type).

1.   Open the "Attribute of Soils" table.

2.   Right-click on the "MUSYM" field heading and click "Summarize" from the context menu (Figure 176).

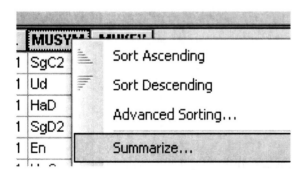

Figure 176: Access Summarize Dialog

3.   The "Summarize" dialog opens (Figure 177). In the "Selective field to summarize" drop-down box click on "MUSYM" field name.

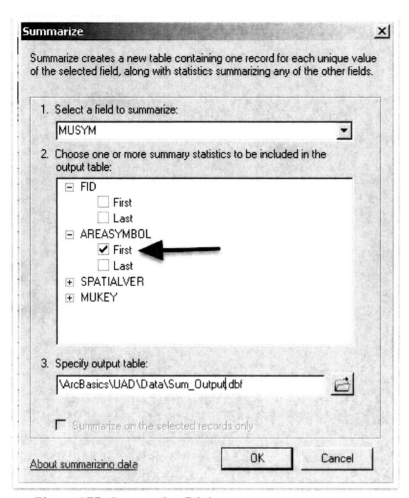

Figure 177: Summarize Dialog

4. In the "Choose one or more summary statistics to be included in the output table:" window, check the box next to the sum under the "AREASYMBOL" field (Figure 177).

5. In the "Specify output table" box, navigate to the "..\ArcBasics\UAD\Data" folder. Click "Save".

6. Click "OK". ArcMap creates a table of the total acreage for each unique value in the "MUSYM" field.

7. Click "Yes" to add the result table to the map.

8. Scroll to the bottom of the TOC and open the new table.

9. Close the summary table when you are finished examining it.

#

Delete an Existing Field and Add a New Field to Table

Using either ArcMap or ArcCatalog, you can add and delete fields in the attribute table. In the next example, we will add and delete a field to the attribute table. In order to add and delete fields, the table does not need to be editable. **Keep in mind that deleting a field cannot be undone.**

1. Open the attribute table for the "Parcels" layer.
2. Right-click on the "Acres" field and click "Delete field" (Figure 178).

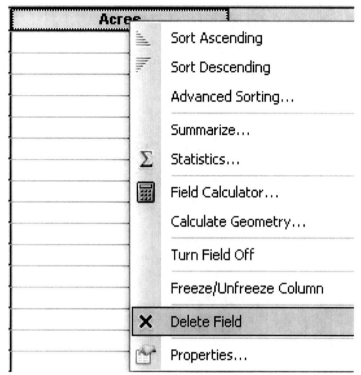

Figure 178: Access Delete Field

3. Click "Yes" in the "Confirm Delete Field" dialog. The "Acres" field is deleted and cannot be undone.
4. In the attribute table, click the "Options" button and click "Add field".

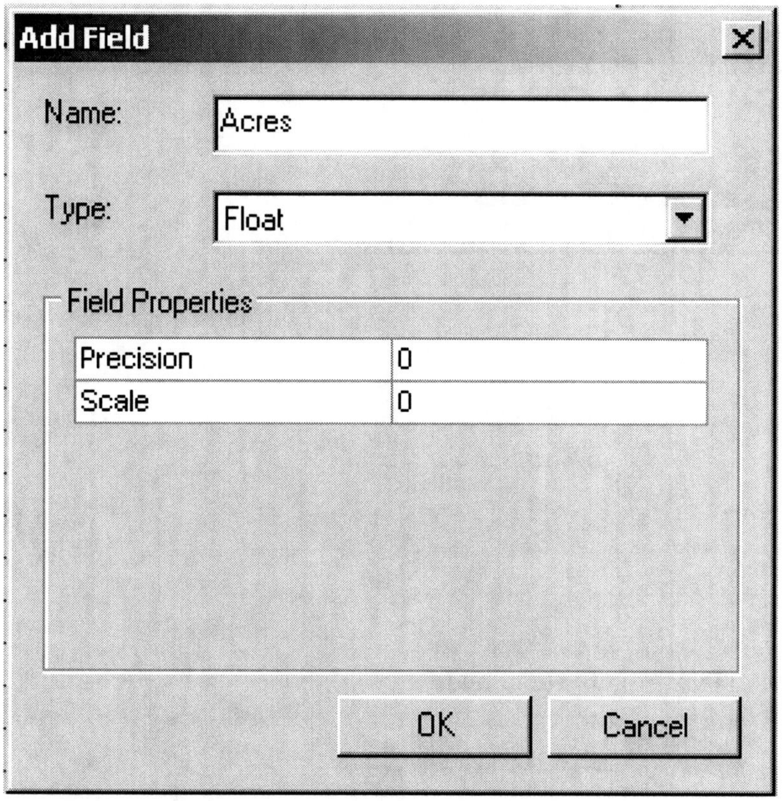

Figure 179: Add Field Dialog

5. Enter the following information in the "Add Field" dialog (Figure 179):

    ☐   Name: Acres
    ☐   Type: Float
    ☐   Field Properties: Use Default Values

6. Click "OK". The acreage field is added to the table (new fields are added to the far right of the attribute table).

                                        #

Calculate Field Values

As you look at the Parcels attribute table, notice the field names are light gray. The colored background denotes that the attribute table is not editable (Figure 180 - 1). When field names are on a white background (Figure 180 - 2), the table is editable. In order to edit the values in a field, the table should be in edit mode. Field values can be calculated outside of an edit session, but it is not recommended. By calculating outside of edit mode, the undo function is not available. Therefore edits cannot be undone.

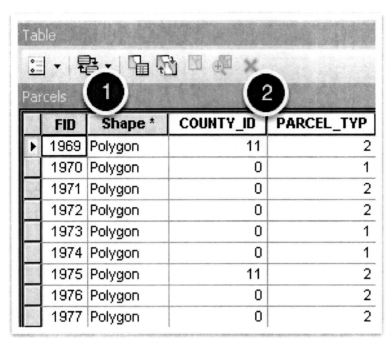

Figure 180: Non-Editable (1) vs. Editable (2) Fields

1.   In the standard toolbar, click the ⬚ ("Editor Toolbar") button. The Editor Toolbar will open. If you save ArcMap with the Editor Toolbar open, it will still be open the next time you open ArcMap.

2.   In the Editor Toolbar, click on the editor drop-down and click "Start Editing" (Figure 181).

Figure 181: Start Editing

3.   In the "Start Editing" dialog, click on the "...\ArcBasics\UAD\SHP" folder to start editing in. When the folder is selected, notice the list of layers available for editing. Click "OK".

4.   Open the "Attributes of parcels" table. Notice that the field name backgrounds are now white, indicating the table is editable.

5.   Right-click on the "Acres" field heading and then click on "Field Calculator" (Figure 182).

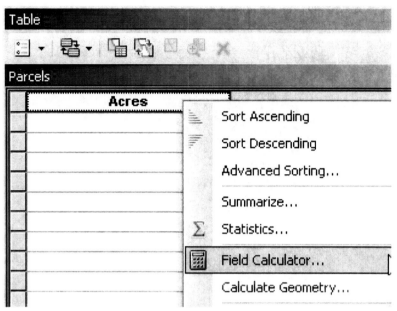

Figure 182: Access Field Calculator

6. When the "Field calculator dialog" opens (Figure 183), scroll down the "Fields" list and click on "Area" field. Verify that the [SHAPE_Area] field is added to the expression box.
7. Click on the division button.
8. In the expression box, enter 4046.9446 after the division symbol. Keep in mind that 1 acre equals 43,560 ft.^{2}.

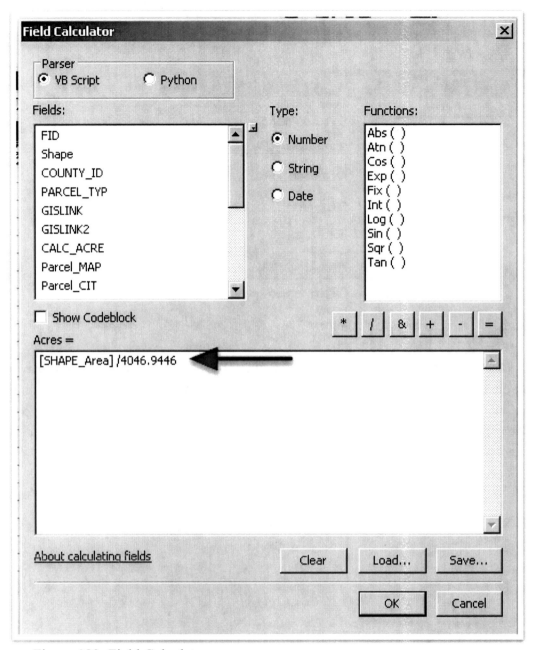

Figure 183: Field Calculator

9. Click "OK" and ArcMap will perform the calculation (Figure 183).
10. In the Editor Toolbar, click the "Editor" drop-down, and click "Stop Editing".
11. Click "Yes" to Save Edits.

#

Table Joins and Relates

As you have seen, there are many operations that can be performed on an attribute tables.
ArcMap also gives you the ability to work with multiple tables simultaneously. The primary

operations are "Join" and "Relate". When working with multiple tables, joins take precedence. If you perform a join after performing relay with the same table, you'll lose the relate relationship and will have to reestablish the join.

In order to decide whether to use join or a relate, you must identify the relationship between the attribute table and to records in the join/relate table. Identify if one or more record(s) in the target table is associated with one or more record in the join/relate table. Identify the fields in the tables that contain the common values upon which the join or relate will be based. The common field values between the two tables may be the same data type but it is not necessary for the common field names to be the same.

#

Table Joins

When tables are joined records from the join table are appended to the attribute table using the common fields. Although the tables appear to be merged, the layout of each table remains unaltered.

When performing a join in ArcGIS, there are two types of relationships. These relationships are:

1. One to one: in this type of relationship, one record in the join table relates to only one record in the target table. For example, when joining the "Parcels" table with a "Structures" table, one record in the parcels table matches each record in the structures table.
2. Many to one: in this type of relationship, many records in the target table relate to one record in the join table. For example, the parcels table contains information about property types. The property type information can be related to a table that contains descriptive information regarding each property type.

#

Join Data from One Table to Another

In the following example, you will join data from a Structures.dbf table to the attribute table for the parcels layer.

1. Open the UAD.mxd located in the "...\ArcBasics\UAD" folder.
2. Click the "Add data" button in the standard toolbar and add the "Structures.dbf" table located in the "...\ArcBasics\UAD\dbf" directory (Figure 184).

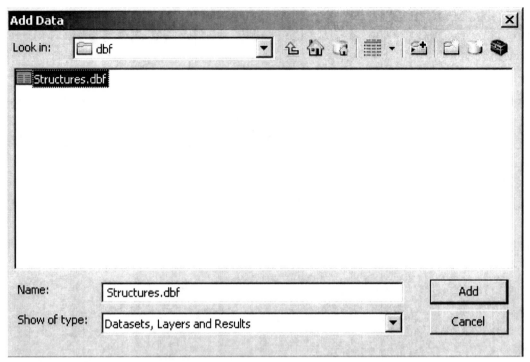

Figure 184: Add Structures.dbf

3. Open the "Structures.dbf" table.
4. Open the "Parcels" table.
5. Take a moment to review that the fields in each table. Notice the "GISLINK" field in both tables.

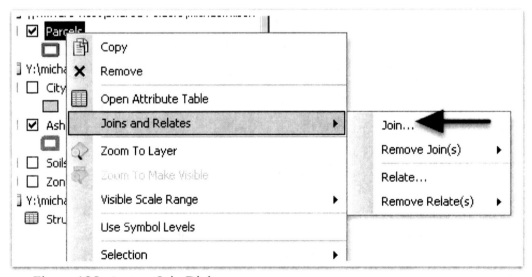

Figure 185: Access Join Dialog

6. In the TOC, Right-click on the Parcels layer and click on the "Joins and relates" option, click on "Join" (Figure 185).

7. In the "Showing data" dialog, in the "What do you want to join to this layer?" drop-down box select "Join attributes from the table".
8. In the "Item 1" drop-down, click on the "Parcels" layer.
9. In the "Item 2" drop down, click on the structures table.
10. In the "Item 3" drop-down, click on the "GISLINK" field. This will create a join based on this field.
11. Under "Join options", click on the option to "Keep all records".

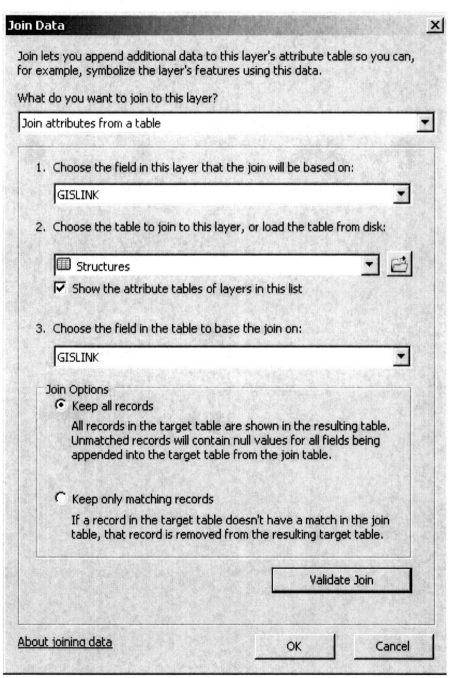

Figure 186: Join Data Dialog

12. Click OK.
13. If prompted to "Create index", click Yes.
14. Close the "Structures" table.
15. The attribute items from the structures table are now joined to the "Parcels" table. Take a moment to review the attribute table.

As noted previously, the join is not permanent. You can save the table with the join data by exporting the table. To export, click the "Table options" button, and choose "Export". Additionally, you can export the data set to create a new layer that will contain the join data.

#

Remove Joined Data

In the following exercise, we know how to remove joined data.

1. In the TOC, right-click on the "Parcels" layer.
2. Click on "Joins and relates", and then click on "Remove Joins" and then "Remove All Joins" (Figure 187).

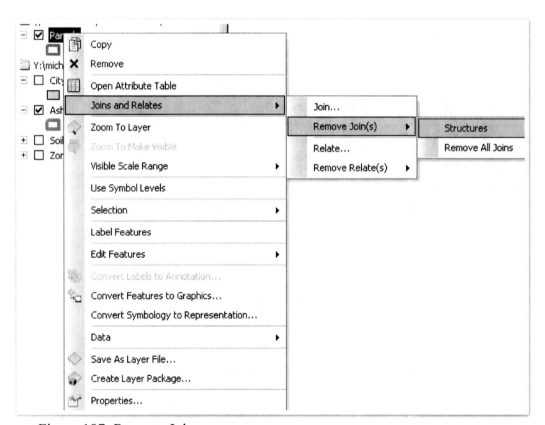

Figure 187: Remove Joins

3. Examine the "Parcels" table and verify that the joint has been removed.

As an alternate to the method outlined above, you can also use the "Layer Properties" and then "Joins and relates" dialog to manage all joins established for the layer or table.

#

Table Relates

When using joins in ArcMap, records from one table are appended to another table. Relates do not append records. Relates create a link between the two tables. To view a relate, you must open the two related tables. You select a record in one of the tables, and the related records are highlighted in the second table. A relate is bidirectional. Therefore, you can select records in either table and the related records will be selected in the other.

As with Joins, there are two types of relate relationships:

1. One to many: each record in the target table is related to at least one record in the relate table. For example, one record in a parcel table may be related to multiple soils records. Basically, a parcel is made up of multiple soil types.
2. Many to many: multiple records in a target table are related to multiple records in the relate table.

#

Relate Data Between Two Tables

In the next example, you will relate data from the Structures.dbf table and the attributes of Parcels layer.

1. Open the "UAD.mxd" project.
2. In the standard toolbar, click the "Add data" button.
3. Navigate to the "Structures.dbf" table located in the "...\ArcBasics\UAD\dbf" folder and add it to the map.
4. Open the "Parcels" table and open the "Structures" table. Review the various fields in each table.
5. Take a look at the "GISLINK" field in both tables. This field will be the basis for the relate.

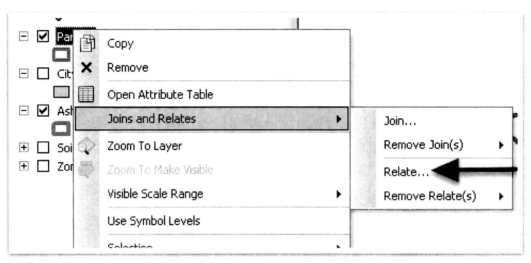

Figure 188: Access Relate Dialog

6. In the TOC, right-click on the "Parcels" layer. Click "Joins and Relates". Click "Relate" Figure 188).

7. In the "Item 1" drop-down box, click on the "GISLINK" field.

8. In the "Item 2" drop-down box, click on the "GISLINK" table.

9. In the "Item 3" drop-down box, click on the "GISLINK" field.

10. In the "Item 4" text box, enter "Soils and Zoning".

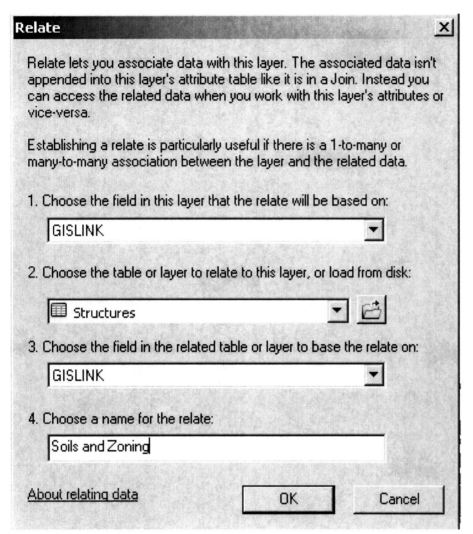

Figure 189: Relate Data Dialog

11. Click "OK". The relate is now established between the two tables (Figure 189).

#

Access Related Records

In the following example you will learn how to access all the data and related table. We will select a record in the parcels layer and looked at all the related records.

1.   In the TOC, right-click on the "Parcels" layer and open its attribute table.
2.   Select a single record in the attribute table.
3.   In the table window, click the "Options" button.

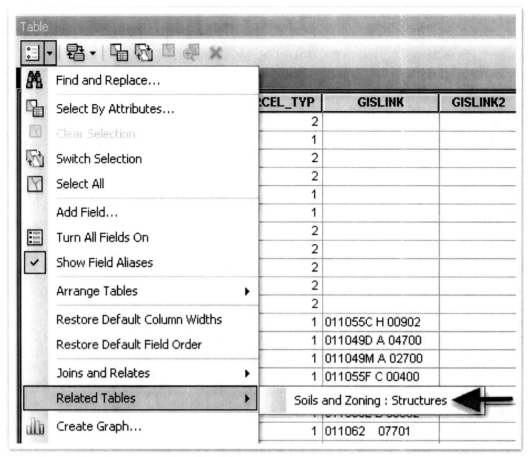

Figure 190: Open a Related Table

4. In the dialog, hover the mouse pointer over "Related tables" and click "Soils and Zoning: Structures" (Figure 190).
5. The "Structures" table will open and the related records will be displayed.
6. Click the title bar of "Structures" table and bring it to the front.
7. Click the "Selected" button. Only the selected records are shown.
8. Take a moment to explore the table and the selected records.

    Keep in mind that the relate works both ways. You can select records in either tables, and then use the method outlined above to display the related records.

                                    #

Remove a Related Table

1. In the TOC, right-click on the "Farms" layer.
2. Click on "Properties" in the context menu.
3. In the "Layer Properties" dialog, click on the "Joins and Relates" Tab.

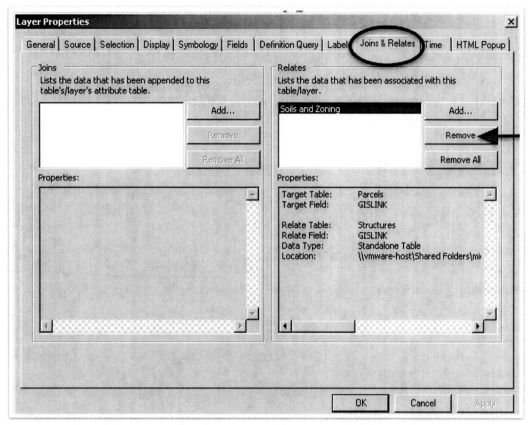

Figure 191: Joins and Relates Tab

4. Click the "Soils and Zoning" relate.
5. Click the "Remove" button (Figure 191).
6. Click "OK". Browse the attribute table and notice that the related table is no longer present.

You can also remove the relate, by right-clicking the layer in the TOC, highlighting "Joins and relates" and clicking "Remove Relates".

7. Close all Open tables.

#

Hyperlinks: Display Images of Spatial Features in a Data Layer

In ArcMap, hyperlinks can be used to provide additional information about features. Hyperlinks can link to documents, URLs, or micros. The links are defined either as field values or dynamically. In the following example, you will use field-based hyperlinks.

We will create links between features and digital images.

1. Open the "UAD2.mxd" file located in the "\ArcBasics\UAD" folder.

2. In the TOC, right-click on the "Robertson County, TN Cities" layer and click properties in the context menu.
3. In the "Layer Properties" dialog, click the "Display" tab (Figure 192).
4. Check the box next to "Support hyperlinks using fields".
5. Use the drop-down list to select the "Link" field as the URL.
6. Click "OK".

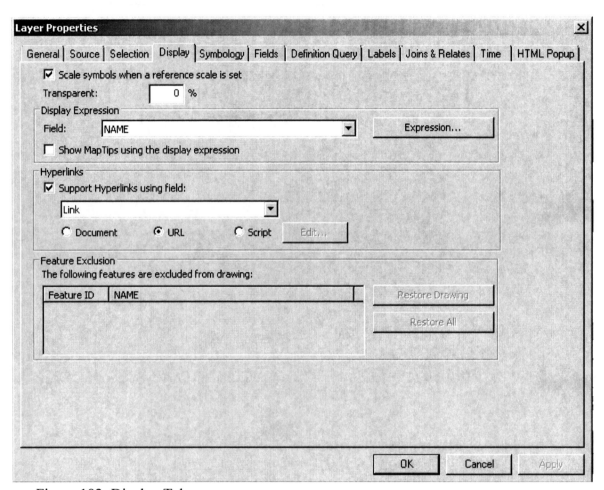

Figure 192: Display Tab

#

Access a Feature's Hyperlink

The following exercise, you will learn how to access the features hyperlink.

1. Use the "Zoom In" tool from the standard toolbar and zoom in to the Cities.

2.   On the "Tools" toolbar, click the ("Hyperlink") tool. Once you click this button, the feature with a hyperlink is outlined in blue. Additionally, the cursor will change to a blackened lightning bolt and a pop-up tip will show the image name.

3.   Click on one of the outlined emergency phone. In new window will open showing the associated URL.

#

Exercise 1

You are a GIS technician for a Chamber of Commerce. An industry is interested in building a new warehouse. You need to create a map showing all the agricultural parcels that are over 150 acres with a land value ("LANDVAL") of >$90,000.

The data for this project are located in the "...\ArcBasics\UAD\Data" folder. You have been given the Parcels layer with includes both the acres ("CALC_ACRE") and the property type ("PT"). The "LANDVAL" is located in the "Land.dbf". Using this data, create your map.

1.   Using the layers and data listed above, create a map project called Industry.mxd and save it to the "...\ArcBasics\UAD\Data" folder.

2.   Add the layers to the map.

3.   Change the Parcels to only display outlines.

4.   Select the parcels that are agricultural parcels that are over 150 acres with a land value of >$90,000.

Chapter 10 - Creating and Editing Data

Creating and Editing Data

In previous sections, we learned how to manipulate existing data. What we have not learned is how to proceed when no data exists. This section will cover how to create and edit data.

ArcMap has powerful tools to allow the user to both create and edit data. As stated earlier, there are three types of vector data: points, lines, and polygons. In this chapter, I will discuss working with the various types of vector data.

#

Objectives:

Creating and editing spatial data is one of the most important tasks for a GIS professional. At the end of this chapter, you should be able to:

- Create a new Layer.
  - ☐ Via ArcCatalog
  - ☐ Via ArcCatalog in ArcMap
- Digitize (heads-up) and edit features
- Digitize Polygon Features
- Edit Vertices and change the shape of a feature
- Add tabular x/y coordinate data to a map document
- Edit data in an attribute table

#

As the name Geographic Information System states, a GIS requires spatial data. In order to be an effective GIS user, you will need to have the ability to create spatial data. The techniques describes in this section are almost entirely universal. In short, if you know how to create environmental GIS layers, you should be able to create infrastructure data. As I often explain to my GIS students, data is data. It can be created in much the same ways but it can be used differently.

Learn to create GIS data and you can start to build your own system.

#

Create a layer via ArcCatalog

1. Open ArcCatalog. Navigate and Select the "\ArcBasics\CED\Data" folder.

2.   Click on the "File Menu" Item.

3.   Mouse over "New" and click "Shapefile" (Figure 193) to open the "Create New Shape File" dialog.

Figure 193: Create New Shapefile...

4.   Name the new shape file "Buildings" and set the "Feature Type" to "Point" (Figure 194).

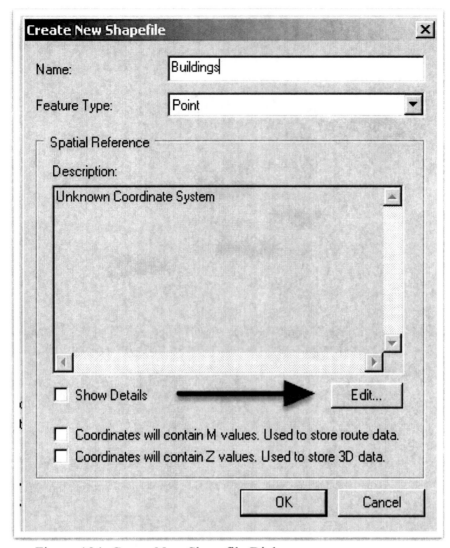

Figure 194: Create New Shapefile Dialog

5. Define the Coordinate System for the shape file. Click the "Edit" button to open the "Spatial Reference Properties" window.

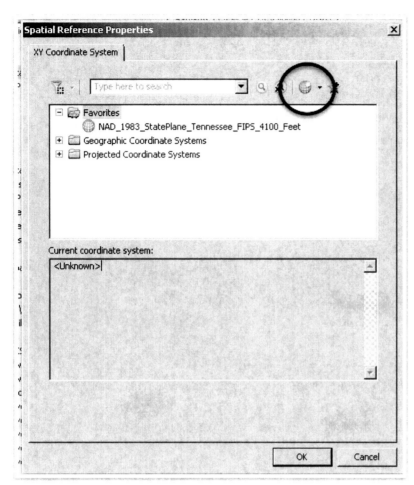

Figure 195: Import Coordinate System

6. Click the dropdown on the "Add Coordinate System" button and the select "Import" (Figure 195).
7. In the "Browse for Data Set" window, navigate to the "...\ArcBasics\CED\SHP" folder and click on the "apsu_buildings.shp".
8. Click "Add" and then click "OK" to close the "Spatial Reference Properties" window.
9. Click the "OK" button to close the "Create New Shape File" window.

\#

Create data via ArcCatalog in ArcMap

1. Open ArcCatalog. Navigate and Select the "\ArcBasics\CED\Data" folder.
2. Click on the "File Menu" Item.
3. Mouse over "New" and click "Shapefile" (Figure 193) to open the "Create New Shape File" dialog.

Figure 193: Create New Shapefile...

4. Name the new shape file "Buildings" and set the "Feature Type" to "Point" (Figure 194).

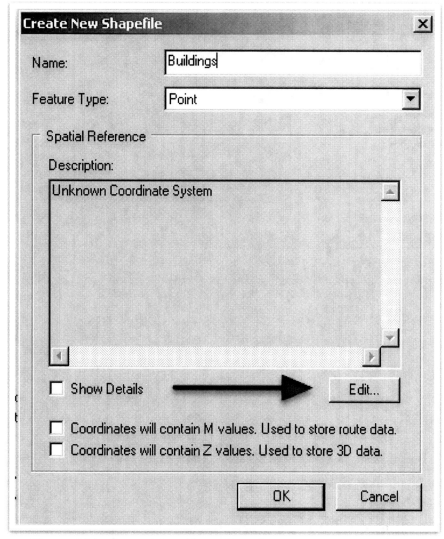

Figure 194: Create New Shapefile Dialog

5. Define the Coordinate System for the shape file. Click the "Edit" button to open the "Spatial Reference Properties" window.

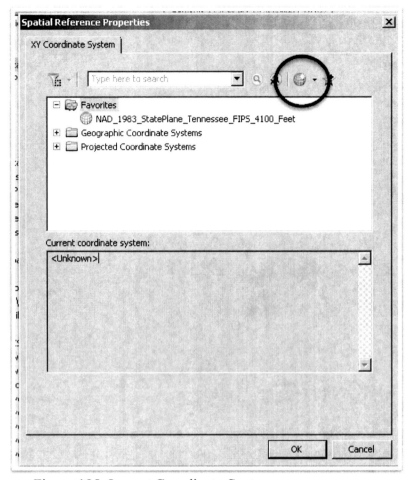

Figure 195: Import Coordinate System

6. Click the dropdown on the "Add Coordinate System" button and the select "Import" (Figure 195).
7. In the "Browse for Data Set" window, navigate to the "...\ArcBasics\CED\SHP" folder and click on the "apsu_buildings.shp".
8. Click "Add" and then click "OK" to close the "Spatial Reference Properties" window.
9. Click the "OK" button to close the "Create New Shape File" window.

#

Heads-Up Digitizing

Digitize Point Features

In GIS, there are numerous methods to create data. One of the most common methods is heads-up digitizing. Heads-up digitizing is the process of adding a base layer to a map and creating the features from the base. For example, a user adds an aerial photo and creates a polygon layer based on the pastures in the image. In the following demonstration, we will digitize features from an aerial photo.

1. Open ArcCatalog. Navigate to the "\ArcBasics\CED\SHP" folder.
2. Open ArcMap, and open the "CED_Buildings.mxd" located in the "...\ArcBasics\CED" folder.
3. Drag the new shape file from ArcCatalog by clicking and holding on the "buildings.shp" and drop it into the TOC in ArcMap. If a warning appears, click "OK".

Figure 181: Start Editing

4. In the "Editor" toolbar, click on the "Editor" drop-down arrow and click "Start Editing" (Figure 181).
5. If the "Starting To Edit In a Different Coordinate System" dialog appears, click the "Start Editing" button.

Note: it is not recommended that you edit the shape or location of spatial features in a layer that are being projected on the fly (covered at the end of book)! None of the layers in this instance are being projected on the fly, so it is safe to edit it even if you receive the warning "Starting to Edit In a Different Coordinate System..." However, it is important to be aware of this in other instances.

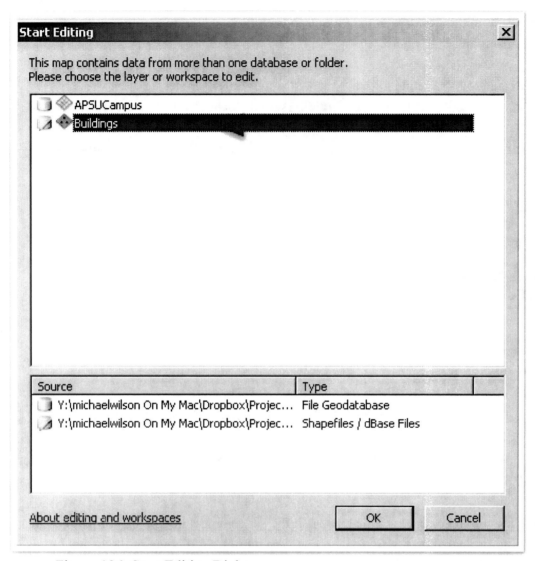

Figure 196: Start Editing Dialog

6. In the "Start Editing" dialog, click on the "Buildings" layer and click "OK" (Figure 196).

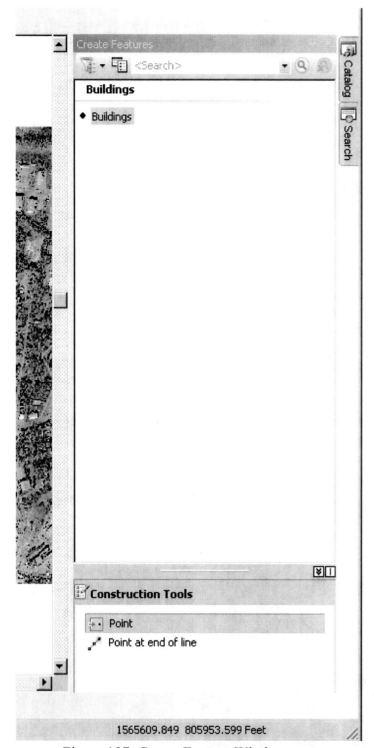

Figure 197: Create Feature Window

7. In the "Create Feature" window, click on "Buildings", and select "Point" in the Construction Tools (Figure 197).

If the "Create Features" window is not visible, go to the Editor toolbar and click "Editing Windows" and "Create Features".

8. Move the cursor to the center of a building and click. A point will be added to represent the building. Repeat this process on several other buildings.
9. In the "Editor" toolbar, Click on the "Editor" drop-down list and click on "Stop Editing" button.
10. When prompted, click "Yes" in the save dialog to save the edits.
11. You can change the point symbol and size by clicking on the point in the TOC.

#

Digitize Polygon Features

As with the digitization of points, the process to create polygon features is similar. In the following exercise, you will learn how to create polygons from the athletic fields on the APSU Campus.

1. Open the " CED_farms.MXD".
2. In ArcCatalog, create a polygon shape file. Name the shapefile "AthleticFacilities.SHP" in the "...\ArcBasics\CED\shp" folder and set the "Feature Type" to "Polygon".
3. Define the coordinate system for the shape file. Click the "Edit" button to open the "Spatial Reference Properties" window.
4. Click the "Import" button.

1.
2. Figure 198: Add Athletics

5. In the "Browse for Data Set" window, navigate to the "...\ArcBasics\CED\Data" folder and click on the "Athletics.shp" (Figure 198).

6. Click "Add" and then click "OK" button to close the "Spatial Reference Properties" window.

7. Click the "OK" to close the "Create New Shape File" window.

8. Drag the "AthleticFacilities" layer into the ArcMap TOC.

9. Begin editing the "AthleticFacilities" layer. Click the "AthleticFacilities" layer in the "Create features" drop-down box.

10. Click on "Organize Templates".

11. Click on the "AthleticFacilities" layer, then click on "Templates" and check the "Fields" box.

12. Click on "Finish" and "Close".

13. From the "Construction Tools", click "Polygon".

14. Find the edge of the field in the imagery in ArcMap. Left click to add vertices and trace the outline of the field. Be sure to continue to left click as you move around the field. Pay particular attention to add more vertices in areas where the field is changing (for example, if a line is curved). More vertices will more accurately show the shape.

15. Notice as you move around the polygon, there is line "Rubberbanded" to the cursor. As you trace the field, you can double-click at any time to complete the polygon.

16. After completing your first polygon, continue to create several additional field polygons.

17. Stop Editing and save the edits.

Note: Take a look at the attribute table for the "Fields" layer. Both the area and the perimeter fields of the layer have not been calculated. You may add fields to the table and calculate area in square feet and perimeter in feet (this is the measurement used by the selected Coordinate System), using the procedures outlined in the section entitled "Update Area and Perimeter Field Values" in the Geoprocessing Section

#

Digitize Line Features by Editing an Existing Shapefile

In the following exercise, you will learn how to edit an existing line file.

1. Open the "CED_APSU.mxd" file located in the "" folder.

2. Zoom in to the center of the campus map.

3. Be sure the "Sidewalks" layer is visible.

Figure 199: Start Sidewalk Edits

4. From the "Editor" toolbar, click "Start Editing" (Figure 199).

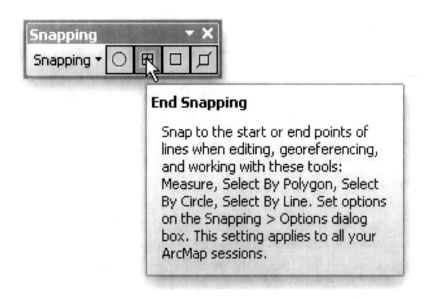

Figure 200: The Snapping Toolbar

5. From the "Editor" drop-down box, click on "Snapping". The snapping toolbar will appear (Figure 200).

   ** Note: the "Snapping" toolbar can be used to establish the location of vertices in relation to other features. It can also be used to move a feature to a precise location in relation to other features.**

6. In the "Snapping" toolbar, click the "end snapping" button. This will enable endpoint snapping (Figure 200).
7. Close the "Snapping" toolbar window.

8. Edit the "Sidewalks" layer by continuing the sidewalk centerlines throughout the center of campus.

Figure 201: Edit Sidewalks

9. Move the cursor and click along the sidewalks (Figure 201).
10. Double-click the left mouse button to stop editing the feature.
11. If prompted, save the edits.

#

Edit Vertices to Change the Shape of a Line

In the following exercise, you will learn how to edit vertices and change the shape of a line.

1. Open the "" file located in the "" folder.

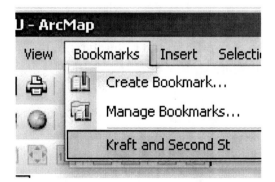

Figure 202: Kraft and Second Bookmark

2. Zoom in to the intersection at Kraft and Second Street (use the Bookmark - Figure 202).
3. From the "Editor" toolbar, click on the "Edit" tool.

Figure 203: Edit Vertice

4. Double-click on the roads at the intersection and align vertices will appear. The vertices will be represented as green squares. The "Edit Vertices" toolbar will appear (Figure 203).

5. Use the "Edit" tool ⬚ to move, add, and remove vertices until the road matches the aerial photography.
6. Click "editor" and click "Stop Editing". Click "Yes" to save edits.

To move a vertex, place the cursor over a vertex, and left-click, hold, and drag the vertex to a new position.

To add a vertex, place the cursor over a line between the vertices, and right-click. In the context menu, click "Insert Vertex" and a new vertex will appear.

To remove the vertex, place the cursor over a vertex and right-click. Click "Delete Vertex" and the text will be deleted.

#

Create a Point Layer from Tabular X, Y Data

In ArcMap, you can take a table with X, Y coordinates and create a point layer. The X, Y coordinates must be formatted into separate fields (i.e., both a latitude and longitude field. The two separate fields must be present. As with any table, additional fields can be present.

1. Open the "waypoints.csv" file located in the "\ArcBasics\CED\Data" folder.
2. Click on "File", click "Add Data", and click on "Add XY Data" (Figure 204).
3. In the "Add XY Data" dialog, navigate to the "\ArcBasics\CED\Data" folder.
4. Click on the "waypoints.csv" file and click "Add".
5. Click on the "X Field" or the "Y Field" dropdowns and select the appropriate X and Y fields.
6. Click "OK".
7. A new "waypoints.csv" layer will appear in the TOC. Right-click on the layer and choose "Zoom to Layer".
8. Click on the point symbol for the "waypoints.csv" layer in the TOC. Use the "Simple selector" to change the point size to 8 and to change the symbol color.
9. Click "OK" to apply the changes.

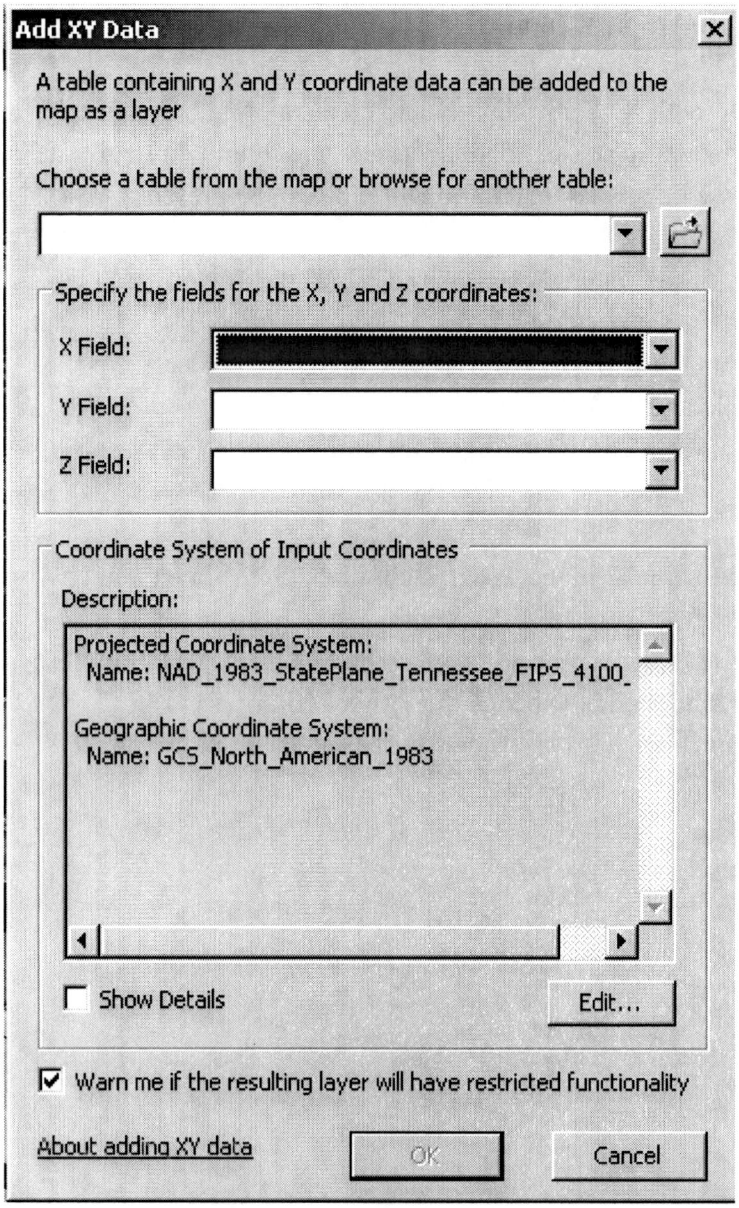

Figure 204: Add X, Y Data

#

Edit Data in an Attribute Table

When you create a layer ArcMap, by default the layer only has three attribute fields. These fields are FID, Shape, and ID. Once a layer is created, you are free to add attribute fields as desired. The same holds true when you have an existing layer and desire to add additional attribute fields. In the following exercise you will learn how to add attributes.

1. Open a new instance of ArcMap.
2. Click the "Add Data" button and navigate to the "\ArcBasics\CED\Data" folder and add the "Buildings" file.

3. Once you have added the layer to ArcMap, right-click on the "buildings" layer in the TOC and click "Open Attribute Table".
4. In the table window, click on "Table Options" and click "Add Field".

Note that while in "Edit" mode, you will not be able to "Add a Field".

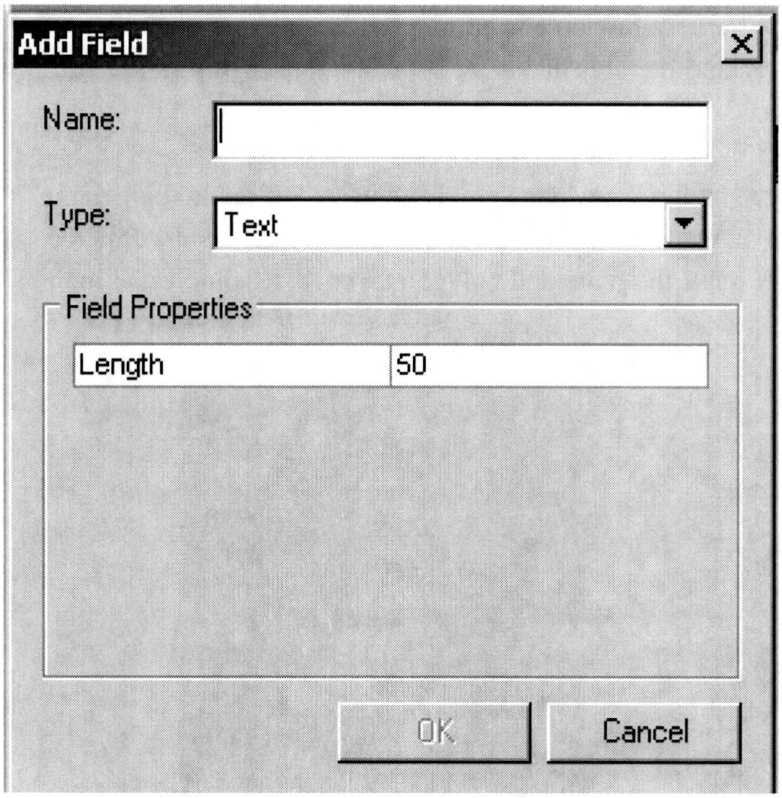

Figure 205: Add Field

5. In the "Add Field" dialog, name the new field "Name".
6. From the "Type" drop-down list, Click "Text".
7. Use a field Length of 30.
8. Click OK.
9. In the "Editor" toolbar, click the "Editor" drop-down and click "Editing".
10. Choose "Buildings" in the "Start Editing" box.
11. Right-click on "Buildings". Click "Table Options" and click "Select All" to select all the records within the table.
12. In the "Editor" toolbar, click on the "Attributes" button. The "Attributes" dialog will open.
13. To add a building name, click in the "Name" field in the attachment table and type in the desired name.
14. Fill in the names of each of the Buildings in the attribute table.
15. Close the "Attributes" dialog when you are finished.

16.   From the "Editor" toolbar, click on "Stop Editing" and then click "Yes" to save your edits.

17.   Close the attribute table.

#

Creating and Editing Data Exercise

Take a little time to explore different drawing and editing functions available in ArcMap. There are several menu items, buttons, and features that were not examined in this module that may be important to you in your work.

Use the APSU Campus image and the knowledge you have just acquired to digitize the parking lots, create a point shapefile of residence buildings, and edit the roads layer within the outlined area shown below. You may reuse the point and polygon layers that you created in this module.

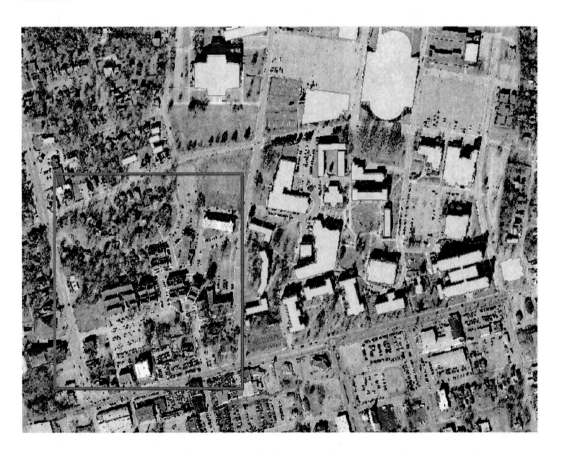

1.   Digitize the parking lots in the outlined area.
2.   Create a "Roads" shapefile and digitize the streets by following the roads in the image.
3.   Using the image as a guide, fix and extend the Sidewalks layer.
4.   Add points to the Buildings layer to represent the buildings in the outlined area.

Chapter 11 - Coordinate system and Projection

Coordinate system and Projection

The power of GIS is tying database information to spatial location. In order to achieve this, database information is projected into a coordinate system. A projection is a model to represent the Earth's surface in a two-dimensional plane. Because the Earth is a sphere mathematical models are needed to transform spherical coordinates such as latitude and longitude to planar coordinates. Planar coordinates can be used to make various measurements including areas.

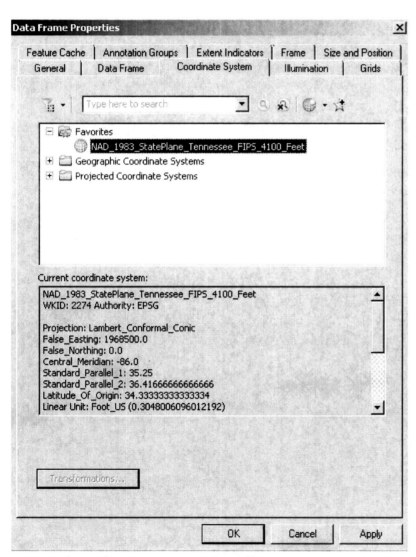

Figure 206: Data Frame Coordinate Systems

A coordinate system is a method to reference the planar coordinate system. There are many different coordinates systems that can be used. Each system has its strengths and weaknesses. Some of the most common coordinate systems are geographic (latitude and longitude) and projected. Projected coordinate systems are those such as UTM and State Plane.

In this section, we will look at projections and coordinate systems in ArcMap (Figure 206).

#

Objectives:

At the conclusion of this chapter, you will be familiar with:

* How to align layers in ArcMap
* Identify a dataset's coordinate system
* Project layers on-the-fly
* Save a layer to a different projection

#

Using Coordinate Systems in the Real World

In order to utilize data in GIS, it needs to be in a coordinate system. The data is referenced to the surface of the Earth using either a geographic coordinate system or a projected coordinate system. Geographic coordinate systems use latitude and longitude for coordinates. Even though only two coordinates are required to locate a point on the Earth's surface, latitude/longitude are three-dimensional coordinates because the Earth's surface is three-dimensional.

Projected coordinate systems use a mathematical conversion to transform latitude and longitude coordinates that fall on the Earth's three-dimensional surface to a flat two-dimensional surface. A projected coordinate system is made up of a spheroid, datum, projection, and horizontal units (i.e. map units).

ArcGIS can work with data stored in either geographic or projected coordinates (Figure 10).

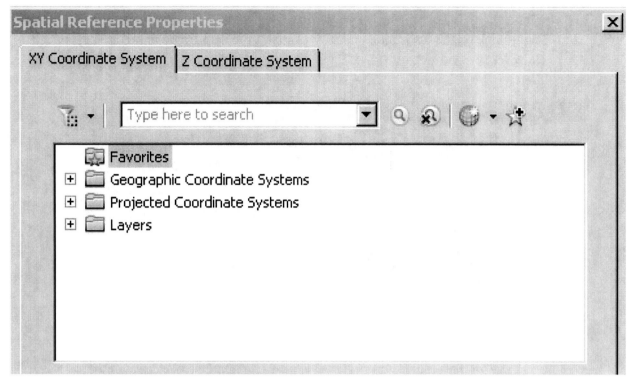

Figure 10: Spatial References Properties Window

For you to make the most of your work with spatial data, always make it a point to know what coordinate system a dataset is in. By knowing this information you will be able to easily manipulate data and make maps that tell a compelling story.

#

How Layers Align in ArcMap

GIS databases store information that contains spatial data. Each data set that contains spatial information stores that data in a coordinate system. As stated previously, because of the irregular shape of the planet, the coordinate system utilized could be latitude and longitude, or any one of hundreds of local coordinate systems. The GIS software recognizes the various coordinate systems and in many cases, performs a transformation to make the layers overlay properly.

ArcMap does excellent job of interpreting the various coordinate systems. Issues arise when either the coordinates system is unidentified or misidentified. Problems can be compounded with the maps data frame. By default, the coordinate system for the data frame is set when the first layer is added to ArcMap. Additional layers are then projected based on the data frame coordinate system. If there is a problem with the data frame coordinates, all layers are transformed incorrectly.

To check the data frame coordinate system:

1. Right-click on the data frame name. By default the data frame name is "Layers".
2. Click on "Properties".

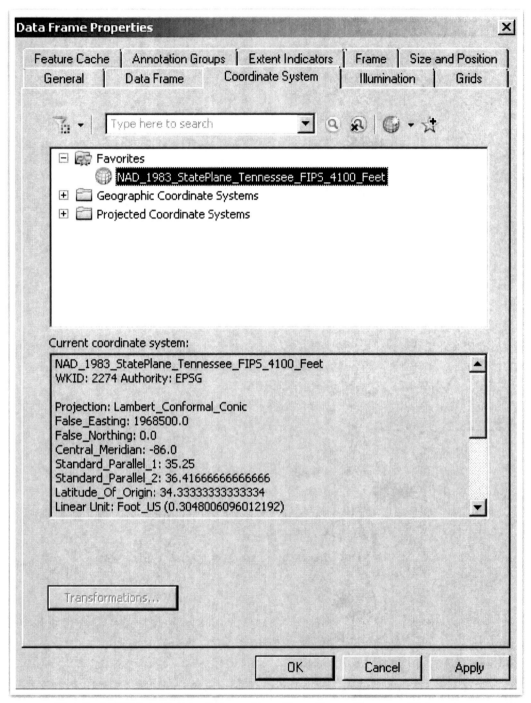

Figure 207: Data Frame Properties - Coordinate Systems Tab

3. In the "Data Frame Properties" dialog, click on the "Coordinate System" Tab (Figure 207).

4. The current coordinate system is listed.
5. To change the data frame coordinate system, simply find the new coordinate system you wish to use click on it.
6. Click "Apply" and click "OK".

To view or change a layer coordinates system:

1. Right-click on the layer name.
2. Click "Properties".
3. Click on the "Source" tab in the "Layer Properties" dialog.
4. Notice the layers coordinate system is listed.
5. To Change the coordinate system, open "ArcToolbox" (covered in detail in the Chapter 12 - Geoprocessing Section).
6. Click on "Data Management Tools" -> "Projections and Transformations" -> "Features" -> "Project" (Figure 208)

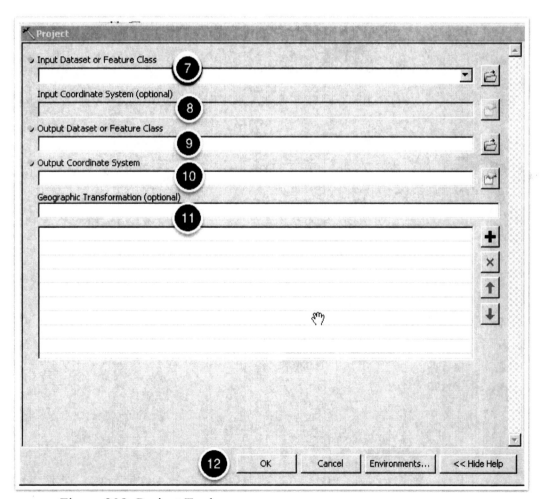

Figure 208: Project Tool

7. Add the Input Dataset.
8. Observe that the current coordinate system will fill in automatically).
9. Select a name and save location for the output dataset.
10. Select the desired coordinate system.
11. Leave the "Geographic Transformation" blank.
12. Click "OK".

#

How To Identify An Unknown Coordinate System

It can be quite tedious to determine the coordinates system of the layer if you do know them in advance. The following methodology will help you to determine if the coordinate system is geographic or projected. Once you make this determination, it is only through trial and error that you will be able to determine the exact coordinate system.

1. Open up a new instance of ArcMap.
2. Click the "Add Data" button in the standard toolbar, add the "apsu_building" file from the "...\ArcBasics\Coord\SHP" folder.
3. Right-click on the layer in the TOC, and click "Properties".
4. In the "Layer Properties" dialog, click on the "Source" tab. Review the extent of the data.

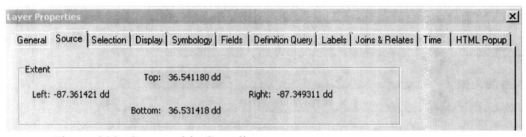

Figure 209: Geographic Coordinates

5. In the "Extent" box, review the values present. If the longitude values are between −180 and 180, and the latitude values are between −190 and 90 you can identify the coordinate system as geographic (Figure 209).
6. If you know that the data is in the United States and that the extent values have values to the left of the decimal are 6, 7, or 8 digits, you can assume the data is a Projected coordinates system such as State Plane or UTM (Figure 210).

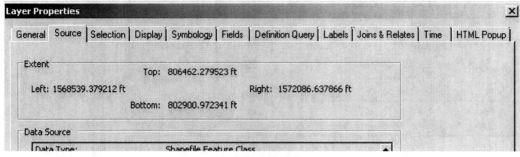

Figure 210: NAD 1983 TN State Plane Feet

For more information about determining coordinates systems, please refer to the ArcMap Help System.

#

Converting a Layer to Another Projection Permanently

In many cases, you will need to convert from one coordinate system to another. This task can be accomplished using ArcMap.

1. Open ArcMap and open up the "Coord" MXD file located in the "...\ArcBasics\Coord" folder.
2. In the TOC, you will see the "APSU_Buildings_DD" layer.
3. Right-click on the layer and click "Properties".
4. In the "Layer Properties" dialog, click on the "Source" tab.
5. Notice that the layer is in "GCS_North_American_1983" coordinates system.
6. Close the "Layer Properties" dialog.
7. In the menu, click "View" and click on "View Properties".
8. In the "View Properties" dialog, click on the "Coordinate System" tab.

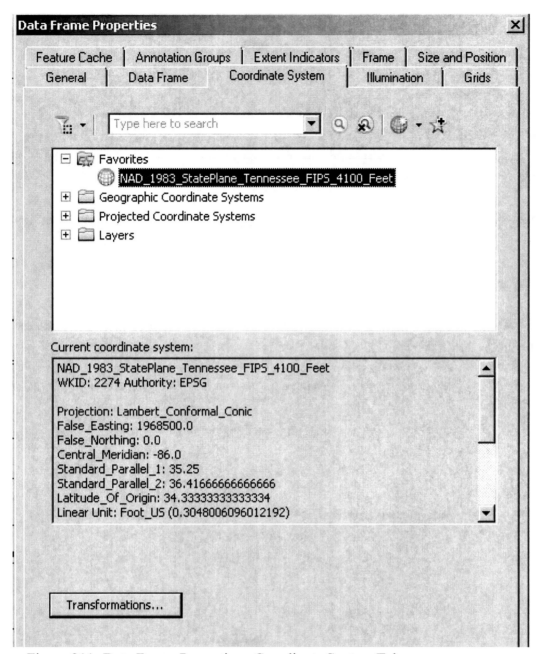

Figure 211: Data Frame Properties - Coordinate System Tab

9. Set the view coordinates system to
    "NAD_1983_StatePlane_Tennessee_FIPS_4100_Feet" (Figure 211).
10. Click "Apply" and click OK to close the dialog.
11. The "APSU_Buildings" layer is now projected on the fly to match the view.
12. Right-click on the layer and click "Save Data".
13. Select a name for the file and click "OK".
14. A dialog will open, notice and click the choice to save the data in the view coordinate
    system.

15. Click "OK" to finalize.
16. When prompted, click "Yes" to add the new layer to the TOC.
17. Check the coordinate system of the new layer and notice that it is now converted to match the view.

#

Project Data On-the-Fly

When ArcMap is combining data in different projections, it has the ability to project data on the fly. Basically, ArcMap interprets the coordinates and performs a transformation so that the data is aligned on the map. This transformation does not alter the data.

1. Open ArcMap and open up the "Coord" MXD file located in the "...\ArcBasics\Coord" folder. Turn off the "APSU_Buildings" layer if present.
2. In the "View" menu, click on "Data Frame Properties".

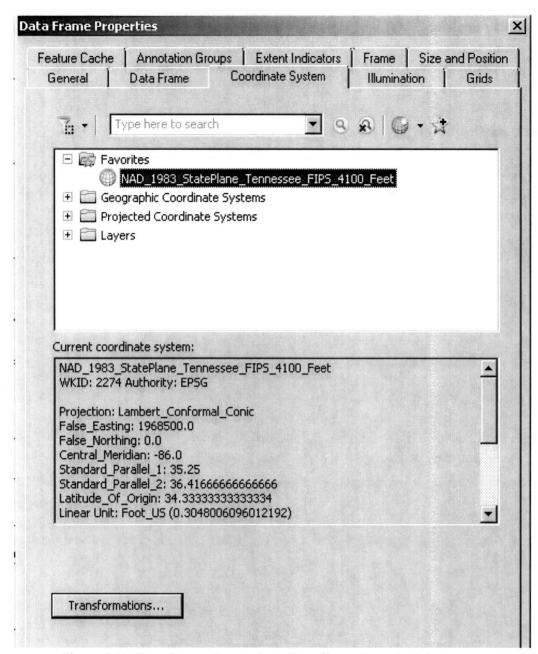

Figure 211: Data Frame Properties - Coordinate System Tab

3. Click on the "Coordinate System" tab. The current coordinate system is set to "NAD_1983_StatePlane_Tennessee_FIPS_4100_Feet" (Figure 211).

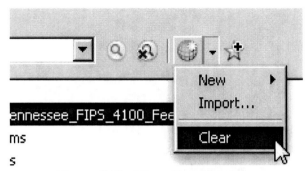

Figure 212: Clear Coordinates

4. In the "Coordinate System" tab, click on the "Clear" button and then click "OK" to close the dialog (Figure 212).

5. Notice that the layers in the map are no longer match. In fact, if you click "Zoom to Extent", you notice the data sets are out in space.

6. In the TOC, Right-click on the "Streets" layer and click "Properties".

7. Once the "Layer Properties" dialog opens, click on the "Source" tab.

8. In the "Data Source" box, notice the coordinate system is defined as "<Undefined>".

9. Click "OK" to close the dialog.

10. In the TOC, right-click on the "apsu_buildings" layer, and click "Properties".

11. Click on the "Coordinate System" tab and define the coordinate system as "NAD_1983_StatePlane_Tennessee_FIPS_4100_Feet".

12. Click "OK" to close the dialog.

13. In the "View" menu, click on "Data Frame Properties".

14. Click on the "Coordinate System" tab. Change the data frame coordinates system to "NAD_1983_StatePlane_Tennessee_FIPS_4100_Feet".

15. Click "OK" to close the dialog.

16. Click on the "Zoom to Map Extent". Notice how all the layers are now align.

17. Close the MXD without saving.

Chapter 12 - Geoprocessing

Geoprocessing

There are numerous geoprocessing tools available in ArcGIS. Users can analyze and combine layers based upon geospatial relationships. The geoprocessing tools are located in ArcToolbox and are available in both ArcMap and ArcCatalog. Some of the tools available include:

- Buffer
- Extraction Functions
- Intersection Functions
- Dissolve
- Merging Data
- Coordinate System Management
- Update Area and Perimeter Field Values
- Model Builder

Each of these tools is essential to performing work with GIS data. ArcToolbox contains many additional tools that provide many advanced functions. These tools can also be combined and edited to create powerful user defined functionality with Model Builder.

\#

Objectives

In this Chapter you will learn the following:

- How to select data based on proximity
- Merging datasets
- Clipping layers
- Updating Attributes such as length, area, and x/y
- Create a Centroid Layer
- Spatial Join
- Create a Summary Table
- Performing a merge, intersect, and dissolve
- An Intro to Model Builder

\#

ArcGIS 10.1 or 10.2 Tool Locations

Clip tool: ArcTool Box- Analysis Tools – Extract - Clip
Erase tool: ArcTool Box – Analysis Tools – Overlay - Erase
Intersect tool: ArcTool Box – Analysis Tools – Overlay – Intersect
Union tool: ArcTool Box - Analysis Tools – Overlay - Union
Dissolve tool: ArcTool Box – Data Management Tools – Generalization - Dissolve
Buffer tool: ArcTool Box - Analysis Tools – Proximity - Buffer
Append tool: ArcTool Box – Data Management Tools-General – Append
Merge tool: ArcTool Box – Data Management Tools- General – Merge

#

Buffer a Point Layer

There are many instances when performing spatial analysis that you may want to create buffers around a layer. In the following example, we will create a map of the lighting on the APSU campus and locate areas that need additional illumination.

1. Open the "GeoProcess.mxd" located in the "...\ArcBasics\GeoP" folder.

2. Click on the ArcToolbox ( ) tool located on the Standard toolbar.
3. Click on the "Analysis Tools".
4. Click on "Proximity".

Figure 213: ArcToolbox - Buffer

5. Double-click on "Buffer" (Figure 213). The "Buffer" wizard dialog will open.
6. In the Input Features window, you can either:

☐ Use the dropdown arrow to choose an existing shapefile that is already loaded into the Map (in the TOC) and double-click.

☐ Using the "Look" button, search for the desired shapefile and click "Add".

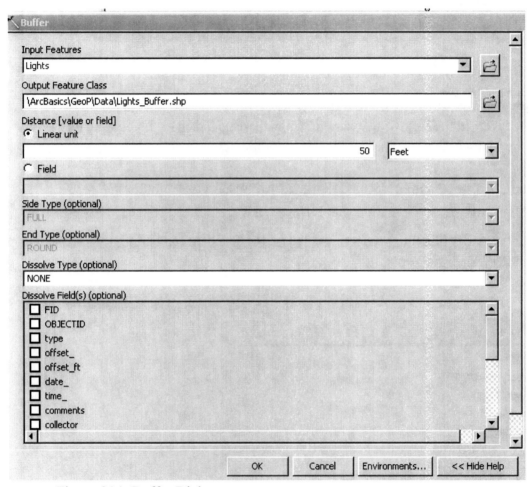

Figure 214: Buffer Dialog

7. For this example, use the dropdown arrow to choose the "Lights" shapefile and double-click (Figure 214).
8. In the "Output Feature Class" text box, save the results as
"...\ArcBasics\GeoP\Data\Lights_Buffer.shp" .
9. Under the Distance window, click the Linear Unit radio button and type in the distance (50 feet) to be buffered.
10. Use the drop down arrow to select "feet" as the units of measurement.
11. Click "OK".

12. Buffers layer will show up on the existing map as a new shapefile in the TOC.

#

Buffering Lines

1. Open the "GeoProcess.mxd" located in the "...\ArcBasics\GeoP" folder.
2. Click on the ArcToolbox tool located on the Standard toolbar.
3. Click on the "Analysis Tools".

1. Click on "Proximity".

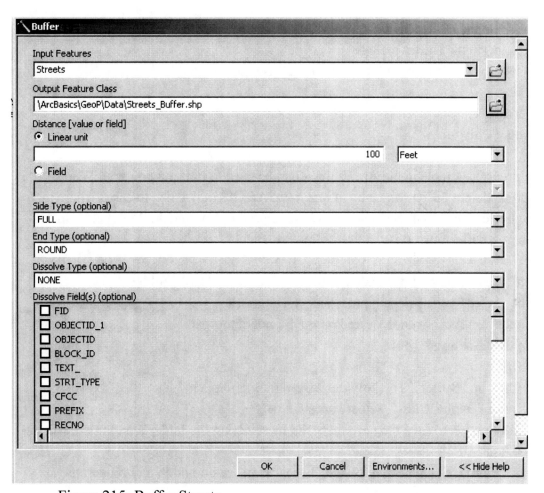

Figure 215: Buffer Streets

4. Double-click on "Buffer" to allow the wizard dialog to open (Figure 215).
5. Under "Input Features Class", select "Streets".
6. Select a destination to save the new shapefile by selecting the folder icon under "Output Feature Class". Set the output to "..\ArcBasics\GeoP\Data\Streets_Buffer.shp".
7. Under the "Distance" window, click the "Linear Unit" radio button and type in the distance (100 feet) to be buffered.

☐   To buffer features of only a select attribute, click the radial button next to "Field" and select the attribute from the dropdown menu.

8. Optionally, from the "Dissolve Type" dropdown menu, the options are:
    ☐   None - will not dissolve the buffers
    ☐   All - each buffer will be merged into one feature
    ☐   List - merges buffers that have the same attributes that are selected in the "Dissolve Fields" window.
    ☐   Leave the default value and select "None".

9. Click "OK" in the Buffer dialog.
10. Click "Close" in the Buffer dialog.
11. Buffers layer will show up on the existing map as a new shapefile in the TOC.

#

Buffering a Polygon

The geoprocessing tools in the ArcToolbox can be used to buffer a polygon. When buffering polygons, do not buffer too many at one time. ArcGIS may crash in cases where to many polygons are selected simultaneously. It is a good practice to save the MXD prior to running the polygon buffer functions.

1.   Open the "GeoProcess.mxd" located in the "...\ArcBasics\GeoP" folder.
2.   Using the "Select" tool, select approximately 10 parcels.
3.   Click on the ArcToolbox tool located on the Standard toolbar.
4.   Click on the "Analysis" Tools.
5.   Click on "Proximity".
6.   Double-click on "Buffer". To allow the wizard dialog to open.
7.   Under "Input Features Class", select Parcels.
8.   Select a destination to save the new shapefile by selecting the folder icon under "Output Feature Class".
9.   Under the "Distance" window, click the "Linear Unit" radio button and type in the distance (50 feet) to be buffered.
10.   Click "OK" in the Buffer dialog.
11. Click "Close" in the Buffer dialog.
12. Buffers layer will show up on the existing map as a new shapefile in the TOC.

#

Merging Datasets

As you are managing datasets in ArcGIS, there are many cases where it would be ideal to merge datasets to simplify data management. In the following exercise, you we learn how to merge several datasets.

1.   Open the "GeoProcess2.mxd" located in the "...\ArcBasics\GeoP" folder.
2.   Click on the ArcToolbox tool located on the Standard toolbar.
3.   Click on "Data Management Tools"
4.   Click on "General".

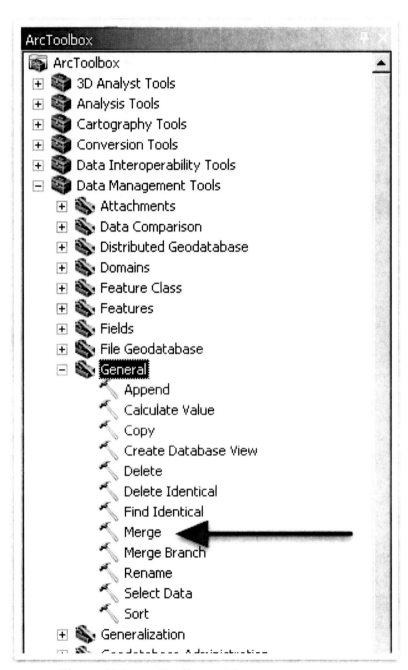

Figure 216: ArcToolbox - Merge

5. Double-click on Merge (Figure 216). The "Merge" Dialog will open (Figure 217).

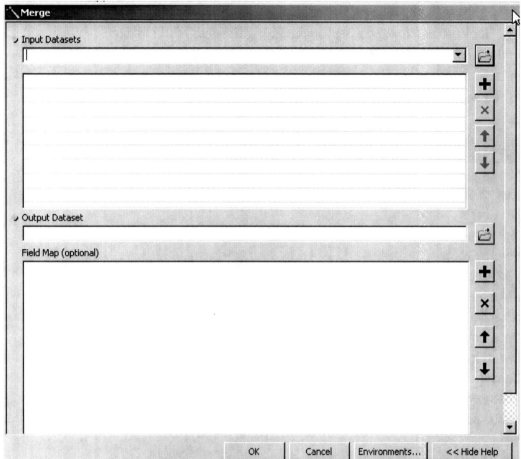

Figure 217: Merge Dialog

6. For Input datasets, select the Robertson Cities datasets you would like to merge from the drop-down box.

- ☐   Ridgetop
- ☐   Greenbrier
- ☐   Adams
- ☐   Portland
- ☐   Coopertown
- ☐   Millersville
- ☐   Springfield
- ☐   Cedar Hill
- ☐   Orlinda
- ☐   White House

☐   Cross Plains

7. For Output Dataset, Browse to "...\ArcBasics\GeoP\Data" to save the output and give name it "RobertsonCities.shp".
8. The fields in each separate dataset are identical. The "Field Map" options are not required inputs. The fields are populated automatically and can be added, renamed, or deleted. The fields and field contents are chosen from the input datasets.
9. Click "OK".
10. The "RobertsonCities.shp" is added to the TOC.
11. Close the map.

#

Clip a Layer

As you are using GIS data, there will be many cases where you are using large datasets. For studies looking at small areas, this may be cumbersome. In instances such as this, you may desire to clip the data down to a smaller area of interest. This task can easily be accomplished using the "Clip" tools in ArcToolbox. In the following example, we will clip the Robertson County Street files down to only include the Streets inside the Springfield City limits.

1. Open the "GeoProcess2.mxd" located in the "...\ArcBasics\GeoP" folder.
2. Click on the ArcToolbox tool located on the Standard toolbar.
3. Click on the "Analysis Tools".
4. Click on "Extract".

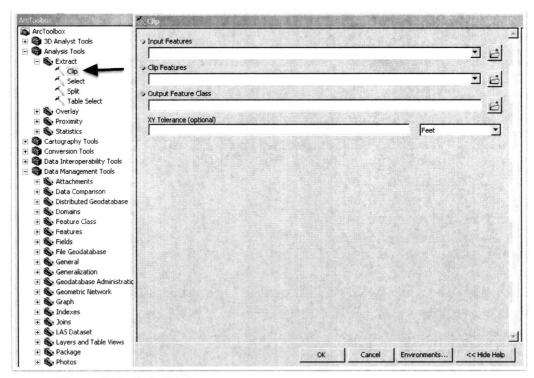

Figure 218: Clip Dialog

5. Double-click on "Clip" allow the wizard dialog to open (Figure 218).
6. From the Input Features drop-down list, select "RobertsonStreets".
7. From the Clip Features dropdown list select "Springfield".
8. In the Output Feature Class text-box, navigate to the "...\ArcBasics\GeoP\Data" folder, enter "SpringfieldStreets.shp" as the file name.
9. Click "Save".
10. Click OK and then Close. The SpringfieldStreets dataset is added to the map. The attributes of SpringfieldStreets are the same as the original "Streets" file.
11. Save the MXD and proceed to the next lesson.

#

Dissolve

The Dissolve Tool aggregates features based upon a specified feature. For example, a particular stream in a rivers layer maybe comprised of many different line segments. The dissolve tool will allow you to combine features into a single segment based upon the stream name.

1. Open the "GeoProcess2.mxd" located in the "...\ArcBasics\GeoP" folder.
2. Open ArcToolbox.
3. Click on click "Data Management Tools".
4. Click on "Generalization".

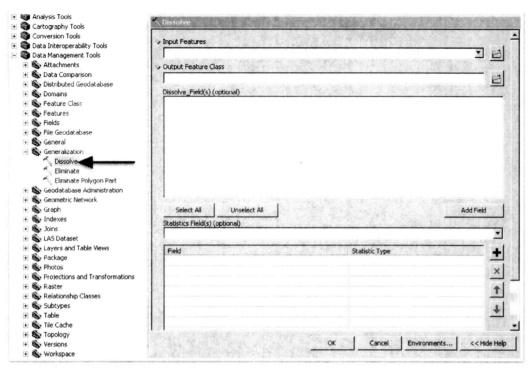

Figure 219: Dissolve Dialog

5. Double click on '"Dissolve" to allow the dialog to open (Figure 219).
6. In the "Input Layer" dropdown, select "Zoning".
7. Name the output layer "DisolveZoning".
8. Leave all other settings as the default values.
9. Click "OK" and "Close". The layer is added automatically.
10. Compare the "Zones" and "Zoning" layers and notice the differences.

#

Union

The "Union" tool can be used to determine how much of one layer is in another layer. For example, we can use "Union" to determine how much of each soil type is in each parcel.

1. Open the "GeoProcess2.mxd" located in the "...\ArcBasics\GeoP" folder.
2. Open ArcToolbox.
3. Click on click "Analysis Tools".
4. Click on "Overlay".

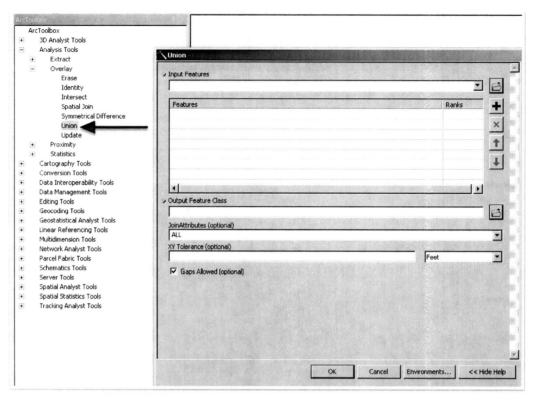

Figure 220: Union Dialog

5. Double click on '"Union". The "Union" dialog will open (Figure 220).

6. Use the "RobertsonParcels" and "USDA_Soils" layers for input.

7. Set the output layer as "ParcelSoils".

8. Click "OK" and "Apply".

9. The "ParcelSoils" is added to the TOC. Notice that "ParcelSoils" contain the attributes for both the "RobertsonParcels" and "USDA_Soils" layers. Keep in mind the both the "Area" and "Perimeter" fields did not update.

#

Intersect

To determine which streams flow through the various cities in Robertson County, you can use the Intersect operation.

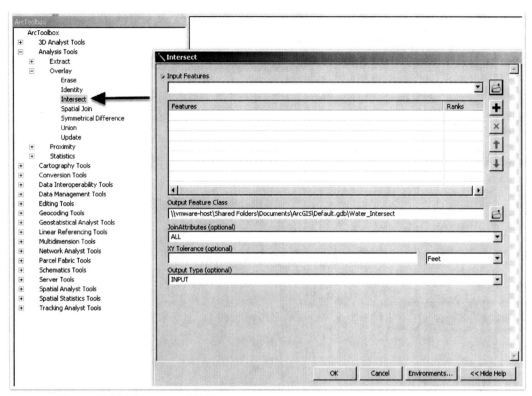

Figure 221: Intersect Dialog

1. In the ArcToolbox window, select Analysis Tools > Overlay, double-click on Intersect to open the Intersect Dialog (Figure 221).
2. Choose "Water" as the first layer and "RobertsonCities" (created earlier) as the second layer. Call the output layer "CityStreams". Explore what the remaining options do, but leave them at the default settings.
3. Click OK and Close.
4. Note that the resulting layer, "CityStreams", appears spatially similar to "Water", but it now has the attributes of "RobertsonCities" layer. The principal difference between the intersect and clip operations is that the latter does not keep the clip layer's attributes.
5. Save the map.

#

Update fields such as Length, Area, Perimeter and X, Y

In the previous lesson, we clipped a Robertson County streets layer to the Springfield City limits. If you look carefully at the attributes of the "SpringfieldStreets" layer and compare them to the "RobertsonStreets" layer you will notice that the "Length" field did not automatically update when you clipped the layer. In the following exercise you will see how to update the "Length" field.

1. Open the attribute table for the "SpringfieldStreets" layer.

2. Click the "Editor Toolbar" button.
3. From the Editor Toolbar drop-down list select "Start Editing".
4. In the "Attributes of SpringfieldStreets" table, right-click on the "Length" field heading to select it.

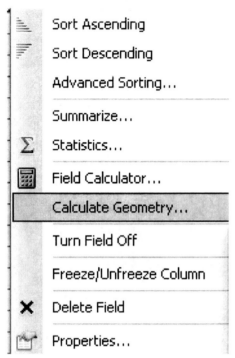

Figure 222: Calculate Geometry

5. Click "Calculate Geometry" (Figure 222). The "Calculate Geometry" dialog will open.
6. Review the default settings.
7. Click "OK" to update the "Length" field.
8. From the Editor Toolbar drop-down list, click "Stop Editing".
9. When prompted, chick "Yes" to save the edits.
10. Close the "Attributes of SpringfieldStreets" table.
11. Save the MXD.

#

Centroids in ArcGIS Basic

1. Open the Attribute table for the "RobertsonCities" layer.
2. In the attribute table, under "Options" choose "Add Field".

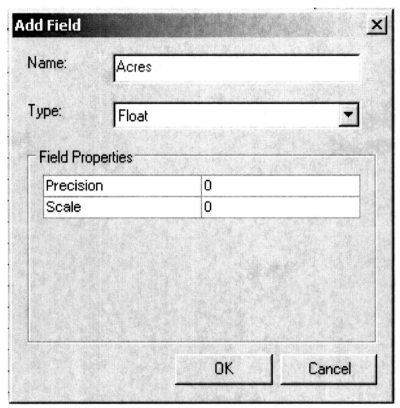

Figure 223: Add Field Dialog

3. Add an X and Y field of type Double (Figure 223).
4. Right-click on the New "X" field and choose "Field Calculator" (Figure 224).

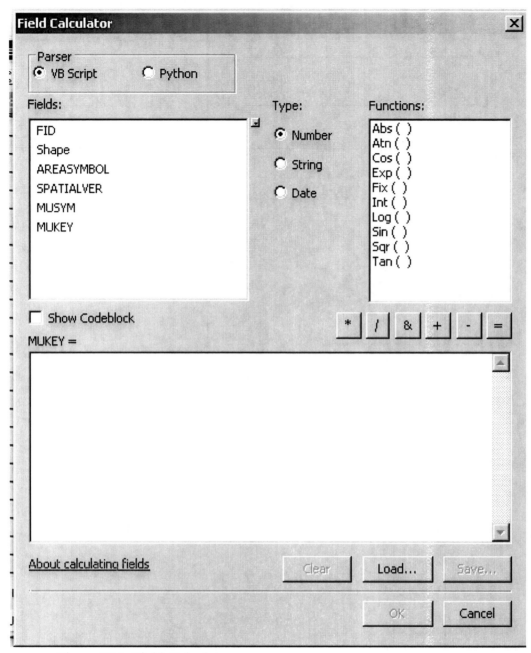

Figure 224: Field Calculator

5. Copy the following code:

```
Dim Output As Double
Dim pArea As IArea
Set pArea = [Shape]
Output = pArea.Centroid.X
```

6. Paste the code into "Field Calculator".

7. Click "OK".
8. Right-click on the New "Y" field and choose "Field Calculator" (Figure 224).
9. Copy the following code:

```
Dim Output As Double
Dim pArea As IArea
Set pArea = [Shape]
Output = pArea.Centroid.Y
```

10. Paste the code into "Field Calculator".
11. Click "OK".
12. You now have the X,Y coordinates of the polygon centroids.
13. To create a separate point layer, copy the DBF file for the polygon layer to a new file and add as a point layer using the "Add X,Y Data..." dialog under the Tools menu.

#

Create a Centroid in ArcGIS Intermediate or Advanced

When working with polygon files, there may be cases in which it is better to represent the polygons as centroids rather than polygons. Centroids are a point that represents the mathematical center of a polygon.

ArcGIS Intermediate or Advanced

1. Open the "GeoProcess2.mxd" located in the "...\ArcBasics\GeoP" folder.
2. Open ArcToolbox.
3. Click on "Data Management Tool".
4. Click on "Features".

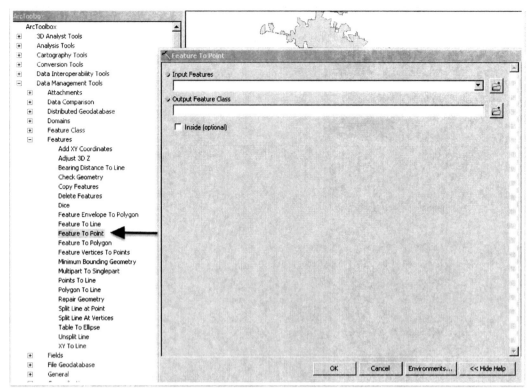

Figure 225: Feature to Point Dialog

5. Click on "Feature to Point" to open the Dialog (Figure 225).
6. Select the "Input Features", use the "Orlinda Zoning" layer.
7. Name the "Output Features" as "OrlindaZoningPoints" and save it in the "...\ArcBasics\GeoP\Data" folder.
8. Specify the type of output point.

☐ CENTROID --Uses the representative center of an input feature as its output point location. This is the default. This point location may not always be contained by the input feature.

☐ INSIDE --Uses a location contained by an input feature as its output point location.

9. Click "Apply". To generate the points.
10. The output layer is added to the TOC.

\#

Spatial Join

A spatial join appends the attributes of one layer to the features in another, based on the spatial relationship between the layers. For example, the relationship between overlapping layers or layers that are close to each other.

1. Open the "GeoProcess2.mxd" located in the "...\ArcBasics\GeoP" folder.
2. Open ArcToolbox.
3. Click on click "Analysis Tools".
4. Click on "Overlay".

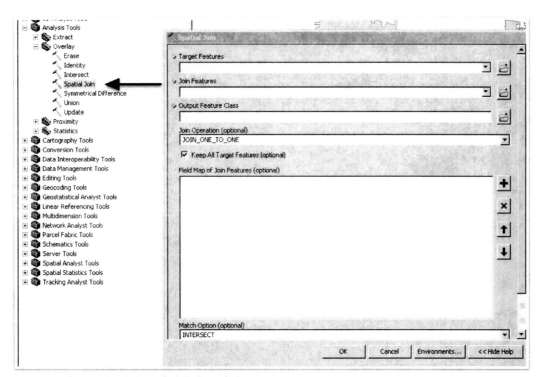

Figure 226: Spatial Join Dialog

5. Double click on "'Spatial Join" to allow the dialog to open (Figure 226).
6. In the "Target Features" dropdown, click "RobertsonSchools".
7. Select "RobertsonParcels" in the "Join Features" dropdown.
8. Save the "Output Feature Class" to the "...\ArcBasics\GeoP\Data" folder.
9. Name the output as "SchoolParcel".
10. Leave all other options as the default values.
11. Click "OK".
12. Upon completion, the "SchoolParcel" output will appear in the TOC.

#

Create a Summary Table

1. Open the "GeoProcess2.mxd" located in the "...\ArcBasics\GeoP" folder.
2. Open the attribute table for the "RobertsonParcels" layer.

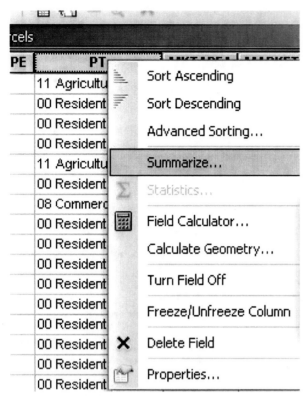

Figure 227: Summarize Menu Option

3. Right-click on the "PT" field heading and click Summarize from the menu options (Figure 227) to open the "Summarize" dialog (Figure 228).

4. Use the "Select a field to Summarize" drop-down, select "PT".

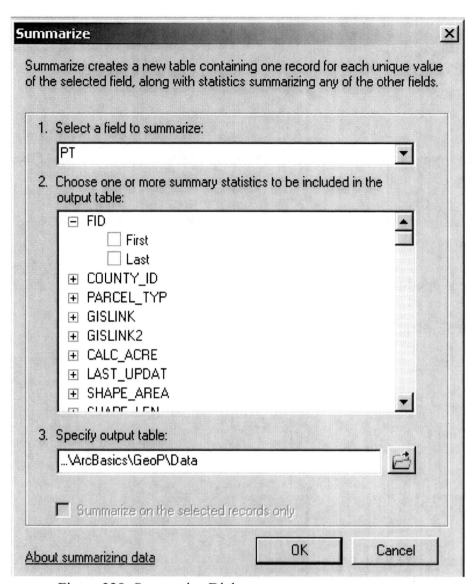

Figure 228: Summarize Dialog

5. Choose the summary statistics to be included in the output table and click on "CALC_ACRE", check the box next to Sum.

6. In the "Specify output table" text box, save the file to "...\ArcBasics\GeoP\Data".

7. Verify that the "Save as type:" drop-down is set to dBase Table.

8. Click "OK".

9. Click "Yes" in the "Do you want to add the result table in the map?" Dialog.

10. Close the "RobertsonParcels" table.

11. Open the analysis table by right clicking on it in the TOC and selecting Open.

12. Examine the contents of the summary table.

13. Close the summary table.

Chapter 13 - Final Exercise

Final Exercise:

Hint: Before starting your analysis, read through the entire exercise and make a list of the parameters and appropriate analytical procedures for solving the problem.

Planning Note: Use the Project planning worksheet on the following page to create a plan for completion of this project.

#

Project Planning Worksheet

 <PROJECT NAME>Project Worksheet

Define Project

Identify Next Action

Project Organization

Project Assistance

Review

#

The Problem

You are the GIS analyst for your County Planning Office. The county was recently awarded a grant to build an additional fire station. This is particularly important because the more citizens within a five-mile radius of a station, the lower the insurance rate for both the homeowner and for the overall community. Lower insurance rates would represent a significant savings for the County's residence.

Your job is to identify possible building locations for a new station. Once you have identified three possible sites, you will need to generate required statistics for each location.

#

Parameters and Objectives

After consulting with the county fire chief and the planner, you have developed the following list of parameters.

You need to identify:
- Current locations of fire stations
- Areas outside of a 5 mile radius of a station
- Approximate population currently within five miles of a station
- 3 Possible station locations to maximize coverage

Your objective is to produce the following:

1. A map showing all of the County's current fire stations based on the X,Y coordinates referenced above. The map will also include parcels, roads, Census data, rivers and five mile buffers. For orientation purposes, add annotation text identifying the roads.
2. A map identical to the above, but showing three possible station locations.
3. A table showing the approximate Current population within five miles of a station, and the three possible locations with population estimates.

#

Available Data

The following spatial data is located in "...\ArcBasics\Final\SHP" folder.
- Stations.xlsx (Excel file of station X,Y data)
- Parcels (parcel.shp)
- Streets (streets.shp)
- Rivers (WaterBodies.shp)

The census **block** data can be added via the US Census website. The data can be obtained via:

http://www.census.gov/cgi-bin/geo/shapefiles2010/main

#

Creating the Map

Use a transparent fill pattern or a layer transparency for any polygons in the final map.

Include all the essential map components, such as a title, legend, data sources, scale bar, neatline, etc.

Glossary

Attribute Table: a table (much like a spreadsheet) that contains information about (and is linked to) spatial features. Each spatial feature has one associated record (row) in the attribute table

Line / Arc: are formed by connecting two X,Y coordinate pairs. When more than two X,Y points are used to create a line, the greater the detail of the line. Lines have length but do not have area. Examples of line data include streets, trails, and railroads.

Metadata: is data about data. GIS metadata describes such things as what real-world GIS features the spatial layer represents, the last update date for the data, the creator (and/or maintainer the spatial data), contact information, and other technical information such as the data layer's spatial reference and storage system.

Point: are geographic features represented by a single X,Y coordinate pair. Points are zero-dimensional features that have neither length or area. Example of point data includes cities, wells, schools, and other points of interest.

Polygon: are an area fully encompassed by a series of connected lines. Polygons are used to represent areas on the Earth's surface. A polygon contains one type of data such as a lake, city boundary, or landmass. Polygons have a measurable perimeter and area.

Raster: In a raster data structure, spatial data are stored in a two dimensional matrix, much like a checkerboard. Each raster, or cell, contains a value.

Vector: In a vector data structure, geographic features (such as wells, roads, national parks, etc.) are represented by points, lines, and polygons that are defined by a set or sets of [X,Y] coordinates.

Vertex: one of a set of ordered X,Y coordinate pairs that define a line or polygon feature.

More simply, a location along a line where the line changes direction giving it shape (similar to a point)

Vector Data File Formats

Shapefile (.shp)

The ESRI Shapefile is a popular geospatial vector data format for GIS data. It is developed and regulated by ESRI. It has become an industry standard for spatial data. Although developed by ESRI, a wide variety of GIS software systems can read and write shapefiles.

A shapefile is comprised of at least 3 core files. The files have the same prefix name and have the following extensions: .shp, .shx, .dbf. There are numerous other optional extensions that may be present. In order to transmit a shapefile to another user, all files with the same prefix name must be sent.

Geodatabase (.mdb and .gdb)

The geodatabase is a collection of spatial datasets and is designed for data storage. It is used by ArcGIS software and managed in either a file folder or a relational database. There are three types of geodatabases: file, personal, and ArcSDE.

Smart Data Compression (SDC)

Smart Data Compression is a compressed GIS dataset format developed by ESRI. It stores all types of feature data and attribute information together as a core data structure.

Layer File (.lyr)

The ESRI layer file does not contain spatial data but does contain reference to symbology and classification data. The layer file is a reference to a dataset.

ArcInfo Coverage

An ESRI ArcInfo Coverage is a georelational data model that stores vector data. The coverage data format is a legacy format.

Raster Data File Formats

MrSID (.sid)

MrSID (pronounced "mister sid") is a proprietary format of developed by [LizardTech](http://www.lizardtech.com/) for encoding of georeferenced raster graphics, such as orthophotos. MrSID is an acronym that stands for multi-resolution seamless image database.

ECW

ECW is a proprietary wavelet compression image format optimized for spatial imagery. It was developed by Earth Resource Mapping (ER Mapper).

JPEG 2000

JPEG 2000 is an image coding system that uses state-of-the-art compression techniques based on wavelet technology. It is a non-proprietary image compression format based on ISO standards, and the standardized filename extension is .jp2.

Based upon information found as:

[ESRI](http://www.esri.com)

[Geospatial Data Formats](http://www.lib.ncsu.edu/gis/formats.html)

Change Log

11/17/2013
- Fixed Table of Contents
- Deleted Extra Whitespace
- Added "Setting up the Data" blurb

12/01/2013
- Fixed sample data URLs

10/4/2014
- Fixed Figure 9 (Borders)

10/19/14
- Cleaned up Mid-Term Exercise
- Fixed Select Data

Endnotes

1. http://en.wikipedia.org/wiki/ArcGIS
2. http://en.wikipedia.org/wiki/ArcGIS
3. http://www.esri.com/software/arcgis/about/gis-for-me
4. http://www.esri.com/software/arcgis/arcgis-for-desktop/extensions
5. http://www.lizardtech.com/
6. http://resources.arcgis.com/en/help/arcgisonline/index.html#//010q00000074000000

CPSIA information can be obtained at www.ICGtesting.com
Printed in the USA
LVOW11s1514300315

432591LV00017B/667/P